THE CHEMISTRY OF LIFE

THE McGRAW-HILL HORIZONS OF SCIENCE SERIES

The Science of Crystals, Françoise Balibar

Universal Constants in Physics, Gilles Cohen-Tannoudji

The Dawn of Meaning, Boris Cyrulnik

The Realm of Molecules, Raymond Daudel

The Power of Mathematics, Moshé Flato

The Gene Civilization, François Gros

Life in the Universe, Jean Heidmann

Our Changing Climate, Robert Kandel

The Language of the Cell, Claude Kordon

The Future of the Sun, Jean-Claude Pecker

How the Brain Evolved, Alain Prochiantz

Our Expanding Universe, Evry Schatzman

Earthquake Prediction, Haroun Tazieff

MARTIN

OLOMUCKI

THE CHEMISTRY OF LIFE

McGraw-Hill, Inc.

New York St. Louis San Francisco Auckland Bogotá
Caracas Lisbon London Madrid Mexico
Milan Montreal New Delhi Paris
San Juan São Paulo Singapore
Sydney Tokyo Toronto

English Language Edition

Translated by Isabel A. Leonard
in collaboration with
The Language Service, Inc.
Poughkeepsie, New York

Typography by AB Typesetting
Poughkeepsie, New York

Library of Congress Cataloging-in-Publication Data

Olomucki, Martin. 1925–
[*La Chimie du vivant*. English]
The Chemistry of Life / Martin Olomucki.
p. cm. — (The McGraw-Hill *Horizons of Science* series)
Translation of: *La Chimie du vivant*.
Includes bibliographical references.
ISBN 0-07-047929-1
1. Biochemistry. I. Title. II. Series
QP514.2.04613 1993
577—dc20 92-14287

The original French language edition of this book
was published as *La Chimie du vivant*, copyright © 1991,
Hachette, Paris, France.
Questions de Science series
Series editor, Dominique Lecourt

TABLE OF CONTENTS

INTRODUCTION

Without a powerful boost from chemistry, the life sciences would never have undergone the revolution which has changed their face forever. The last half century has witnessed many upheavals, but we tend to remember only their technical and social impacts. Our attention becomes focused on genetic engineering, which combines the lessons of molecular biology and the modern theory of heredity to manipulate living things. We rejoice in the promises it holds out for agriculture and medicine; the biotechnologies are attracting investigators and funding, and, since the massive introduction of biotechnology products, the biotech market has given rise to brisk international competition. We are alarmed by the turmoil into which our Western ethical and legal structures have been thrust by genetic engineering and the threat that seems to loom over the future of our species when the human genome becomes open to experimentation.

At the same time, on the same basis, joint research by chemists and biologists is throwing light on the age-old, burning question of the origin of life. This book takes us down the path of this exciting adventure and its seemingly enormous philosophical implications. The reader will discover the reasoning and experiments by which biologists can reconstitute the mechanisms that may have prevailed in forming the first large molecules—the molecules known to have been the repositories of genetic information in every living cell from the very beginning down to the present day.

The work of François Jacob, Jacques Monod, and François Gros at the Institut Pasteur had shown the role of "messenger" ribonucleic acids (messenger RNA) in carrying the information inscribed in deoxyribonucleic acid (DNA) that is the basis for protein synthesis. Then the American biologists Tom Cech and Sidney Altman, whose work earned them the Nobel Prize in Chemistry in 1989, showed that RNA molecules possess properties that were not originally noticed: they can reorganize themselves with no outside help whatever! So there are good reasons to believe that they may have appeared first and originated DNA.

This research, highly technical but described with the greatest clarity in this book, seems to be on its way to solving what the great U.S. specialist in evolution, Ernst Mayr, only ten years ago called "one of the most burning problems facing researchers." At the same time, as often happens at the cutting edge of leading research, it will also answer a philosophical question, very old yet very much alive: what is the origin of life? Contrary to what the positivist caricature of scientific thinking might suggest, the answer will not purely and simply neutralize the question; it will not toss it into a henceforth obsolete metaphysical past. On the contrary, it will invite researchers and philosophers to start looking at this question once again, this time from a new angle.

In *The Origin of the Species*, which he published in 1859, Darwin wrote these words: "There is grandeur in this view of life, with its several powers, having been originally breathed by the Creator into a few forms or into one . . ." Although he wrote in an almost biblical style, Darwin's boldness was considerable: at that time the differences between living organisms were seen as too distinct for them to have a common ancestor. The great British naturalist went even further in a letter dated 1871: "It is often said that the conditions that presided over the production of the first living organisms are still present and must always have been present. But if (and what a big if!) we could imagine that in such a small puddle of tepid water containing all kinds of ammonium and phosphate salts, subject to certain conditions of light, electricity, etc., a protein substance would form, ready to undergo other, more complex changes. In our day, such a substance would immediately be devoured or absorbed, which would not have been the case at a time when no living creature had yet been formed."

This concept, assumed to be purely speculative, might have passed for a rallying cry for the doctrine of "spontaneous generation," at the very time when this doctrine was under fire from Louis Pasteur (1822–1895). In fact, we may say that the question of the origin of life, heavily laden with theological assumptions and wagers, had over the centuries managed to be expressed, in natural history and in biology, only in the terms of this famous doctrine.

Let us not, however, from the vantage point of our superior "knowledge," hastily dismiss this theory as a mere aberration of unenlightened minds: spontaneous generation has actually racked up an impressive number of observations to its credit, besides which it has a very strong internal consistency. Democritus (5th century B.C.) as well as Aristotle (384–322 B.C.) presented it as experimental fact: after all, we can see maggots being born from rotting meat.

But interpretations diverge. For the atomistic "physiologist," observation has confirmed that life is nothing but a natural assemblage of inert materials without the aid of any vital principle or divine intervention. Aristotle, on the other hand, saw in the theory an additional argument supporting his concept of life as the "animation" of matter: a living being is characterized by the innate presence of a vital principle which comprises a "vegetative or nutritive soul" (the faculty of growth and reproduction), a "sensitive or animal soul" (the faculty of feeling, desiring, and being moved), and finally a "reasonable or thinking soul" (the faculty of humanity). Historians argue about whether these are three distinct entities or hierarchized degrees of the same reality in which the inferior can exist without the superior. The fact remains that this Aristotelian concept of life can be applied without difficulty to the story of Genesis and the Jewish concept of the spirit as a "breath" animating the "dust of the Earth" from which humans were formed. The Book of Judges talks about bees engendered by the carcass of a dead lion. Christian tradition followed Aristotle, who wrote in his *History of Animals*: "Bugs come from humor that leaves animals and becomes congealed in the open air. Lice are born of flesh . . . In certain persons, the appearance of lice is in fact a malady originating from the great abundance of moisture in the body."

When Francesco Redi (1626–1697), an Italian gentleman, naturalist, physician, and poet, carried out his huge program of observations on the generation of insects and intestinal worms, what was known as spontaneous (or "equivocal") generation still would not admit to the slightest doubt: it was accepted without discussion that inert matter could give birth to animals of an inferior order: maggots, lice, slugs, wood lice, scorpions, and even frogs or mice.

But when he placed meat under hermetically sealed flasks, Redi saw no maggots emerge, even after several months. At the same time, Father Nicolas Malebranche (1638–1715) was condemning the doctrine that living things emerged from corrupted matter as impious: "I cannot understand why such a large number of persons of good sense could have committed such a gross error. For what is more incomprehensible than that an animal should form from a piece of rotten meat? It is infinitely easier to explain how a piece of rusty iron could change into a perfectly assembled watch, because there are infinitely more springs, and more delicate ones at that, in a mouse than in the most complicated clock." Redi had sown the seed of doubt; Malebranche aggravated it by a philosophical argument.

However, the first systematic uses of the microscope by the Dutch draper Antoni van Leeuwenhoek (1632–1723) reinforced the traditional doctrine. He discovered "living creatures in rainwater that had been standing for several days in a glazed earthenware vase." Very soon, such "animalcules" were discovered in all liquids and, against the opinion of van Leeuwenhoek himself, it was concluded that these "simple" living organisms could be born by the synthesis of inert materials; people even contrived to see this as a sign of supreme wisdom, and it was explained to Malebranche that these animalcules had the providential function of purifying the air. The 18th-century naturalists, anxious to extend the lessons of Newtonian physics to "organized beings" by defending, as did Buffon (1707–1788) and Maupertuis (1698–1759), the idea that living matter is composed of organic "molecules" or "particles," brought extra credit to this doctrine in their great works. Abbé Lazare Spallanzani (1729–1799), a professor in Reggio, and later in Modena and Pavia, conducted experiments from 1765 to 1776 on the origin of the animalcules discovered by van Leeuwenhoek, in response to the experiments of the Irish priest Turberville Needham, who claimed to have effected the genesis of such animalcules from mutton broth. The Abbé believed that he had definitively dismissed the doctrine of spontaneous generation from science.

But in actual fact, as François Jacob showed, "people were not yet ready to renounce the possibility of such a phenomenon." If the question were to be approached scientifically and, in one sense, settled, a new concept of life would have to come along so that it could be formulated in new terms. The turning point came in the middle of the 19th century with the gradual elaboration of the "cell theory" which, thanks to research by the German botanist Matthias Jakob Schleiden (1804–1881) on plants and by the anatomist Theodor Schwann (1810–1882) on animals, designated the cell as the ultimate reality to which the properties of all living organisms could be attributed. From this time on, biologists would scrutinize this "atom of life" and detect the interplay of molecules that enable it to carry out the chemical reactions characteristic of a cell, known as "metabolic" reactions. The pinnacle of cell theory was the 1858 work of the German physician and politician Rudolf Virchow (1821–1902), whose Latin formula is still famous today: *omnis cellula e cellula*—every cell comes from a cell and not from some prior organic magma.

But this theory, long contested by biologists and physicians who, following Haller and Bichat, persisted in viewing fibers as the ultimate element of living things, was unable by itself to impose its most obvious conclusion, that spontaneous generation was a deception.

It was not until some progress had been made in chemical analysis that chemistry, under the guise of the brand-new discipline of organic chemistry, became involved in an area that thus far had been the preserve of naturalists. Even that only became possible after disputes of such acrimony that the winning side became blind to some of the reasons that led to its triumph. The battle waged by the great chemist Louis Pasteur was focused on the extension of his research on fermentation, begun in 1857; he went so far as to say that he was forced into a "digression." His position never wavered: his goal was to prove that microorganisms could not emerge from putrefaction because, on the contrary, they were its agents. This position grew out of a true philosophy of nature, which Pasteur conceived of as being split into two worlds: living things, whose richness, diversity, and organization were already admirably revealed in the vegetable kingdom; and matter, which knew only the homogeneous and the inert.

Pasteur happened to encounter a respected biologist from Rouen (Normandy), Félix Pouchet, author of a book entitled *Hétérogénie ou traité de la génération spontanée* [Heterogeny or treatise on spontaneous generation], published in Paris in 1859. Pouchet reasoned like a biologist and Pasteur like a chemist. The discussion rapidly turned into a quarrel—one of the liveliest and most instructive in contemporary biology. Arguments of authority were soon heard from Pasteur and accusations of atheism were hurled by both sides. Politics soon became involved: was not Pasteur a decorated and honored partisan of Napoleon III? In the end, each party subjected to critical analysis the experiments conducted by the other and reinterpreted them in favor of his own thesis, to the point of the final defeat of "heterogeny." By repeating and refining the experiments of Spallanzani, Pasteur established that, even in microorganisms, living creatures are born only of living creatures.

The irony which is often the salt of political history is not absent from the history of science. By exorcising the demon of "spontaneous generation," Pasteur actually revived that of vitalism, even though this doctrine, which had hampered the constitution of the very organic chemistry that had made his work possible, would now shackle the first steps of "biochemistry," to which it nonetheless opened the door.

We know that vitalism, in its metaphysical and religious form, consists in attributing the specificity of living phenomena to some particular force that escapes the laws of physics and manifests what is alleged to be a design of the Creator. Although the philosophical position of vitalism has been shared, in various forms, by numerous biologists and physicians over the centuries, illustrious chemists such as the Swede Jöns Jakob Berzelius (1779–1848) and the Frenchman Charles Gerhardt (1816–1856) had also made it their own. Marcelin Berthelot (1827–1907), after Jean-Baptiste Dumas (1800–1884) and Michel-Eugène Chevreul (1786–1889), had to fight hard for acceptance of the idea that "the distinction between compounds that form living organisms—organic and inorganic compounds—cannot be considered absolute." And although the idea of a "vital force" was well and truly dead in the chemistry of the 1850s, it remained powerful among biologists.

The German naturalist Ernst Haeckel (1834–1919), an indefatigable proponent of an atheistic "monist" philosophy based on a frankly evolutionist interpretation of the Darwinian theory of natural selection, reflected the general sentiment of his colleagues when he wrote in his *Natürliche Schöpfungsgeschichte* [History of creation]: "Denying spontaneous generation means accepting a miracle: the divine creation of life. Either life appeared spontaneously on the basis of a few specific laws or it was produced by supernatural forces."

The intervention of the theory of evolution, in which Pasteur had shown very little interest, would, however, establish the terms of the problem which today seems to be on its way to being resolved.

After the turn of the century, researchers explored two main pathways simultaneously. The first consisted in stating that life appeared on Earth because it was seeded by "genes" from other worlds. Such was the grandiose theory advanced by the Swedish physicist Svante Arrhenius (1859–1927) under the name of "panspermy" or "panspermia." Based on the discovery of carbon-based elements in meteorites, he considered that living germs could have been carried through interstellar space by cosmic radiation.

The other pathway, exclusively terrestrial to begin with, proved to be highly fertile. In 1924,

Aleksandr I. Oparin, in a short book entitled *The Origin of Life*, broke new ground by pioneering a theory of "molecular evolution" to account for the transition from the nonliving to the living. Oparin blended three essential theses into a single theory and maintained that organic molecules could have evolved outside any organism; they then assembled and formed increasingly complex systems subject to the mechanism of evolution by natural selection.

Martin Olomucki clearly shows how this theory, which is equivalent to having living things arise out of a primitive "prebiotic soup," possibly with the assistance of certain mineral compounds, succeeded in determining the path of research for several decades. In particular, it inspired the famous experiment by Stanley Miller in 1953, whose goal was to reconstitute the atmosphere of the primitive Earth artificially; he sought to show that in such an atmosphere lightning was able to trigger the formation of amino acids. Other elementary materials of life such as nucleic bases and sugars were obtained in similar experiments. In the 1960s, the American Sidney Fox devised a new scenario, presented as a more Darwinian version of Oparin's theory.

Far from confirming such reconstructions, developments in molecular biology actually obscured the question. We learned that, even in the simplest organisms, proteins are assembled from free amino acids according to a genetic program inscribed in nucleic acids. But the link between nucleic acids and proteins is so intimate that, even recently, it was impossible to conceive of one without the other.

The reader will now be able to follow the reasoning which seems to resolve the question by assigning priority to RNA. The conjunction between the theory of evolution and biochemistry, initiated by Oparin and later reinforced by entirely independent research results from thermodynamics and astrophysics, has thus yielded its fruits. The nagging questions contained for millennia in the myths of spontaneous generation have finally led to a precise scientific problem.

The revenge of Pouchet on Pasteur? Certainly not, because contemporary biology dismisses the very idea of "generation" which eventually commingled three very different types of questions: reproduction, heredity, and evolution. And so it goes with the history of creative scientific thought: it does not obey the simplistic pattern drawn by the epistemologists who say that theoretical anticipations are succeeded by experimental confirmations or refutations!

We will see that it is now possible to speak of "chemical evolution" in the Darwinian sense of evolution. Martin Olomucki is squarely in favor of this notion and supplies arguments which, in his view, justify the extension (some will say extrapolation) of the Darwinian scheme to the "prebiotic" world. Let the reader be the judge, and at least understand that, objective as the results of scientific work may appear to be, the elaboration of the hypotheses leading to them always means that in the very process of thinking there is a temptation to assume a philosophical position.

Dominique LECOURT

I

THE BIRTH OF A CHEMISTRY OF LIFE

THE STAGES OF CONVERGENCE BETWEEN CHEMISTRY AND BIOLOGY

These two sciences are relatively new by comparison with, for example, mathematics or astronomy whose origins go back to ancient times. It was not until the 17th century that chemistry began to separate itself from the mystery and secrecy of medieval alchemy and gradually acquired the features of a truly scientific, precise discipline that flourished in the 19th century. Biology is younger still, even though in one sense it has continued the tradition of "natural history." The very name of this "science which studies life" actually did not come into use until 1802 with J.-B. de Lamarck and the German G. R. Treviranus.

These two disciplines are seemingly two distinct scientific fields separated not only by subject matter and purpose but also by the ways in which their adherents think, their experimental approaches, etc. The boundaries are becoming indistinct today, however: it has been found that biological processes are in fact the result of elementary molecular interactions that strictly obey the laws of chemistry, which are themselves subordinate to the laws of thermodynamics. This is a veritable revolution in philosophy.

In fact, many obstacles had to be overcome before we could accept that the difference between a simple chemical reaction involving small molecules and a complex biological phenomenon is not qualitative but merely quantitative.

In the early 19th century, chemists were still categorically postulating that chemical and biological processes were not governed by the same laws. This was a time when they were isolating large numbers of new chemicals, either from inert matter or from living organisms, and believed that substances considered "organic" could not be obtained from mineral compounds. This is why the synthesis of urea, a natural product, from ammonium cyanate, an "inanimate" molecule, accomplished in 1828 by the German chemist Friedrich Wöhler (1800–1882), retrospectively triggered quite a psychological shock. Yet this experiment did not suffice to banish the spellbinding hypothesis of a mysterious "vital force" that governed biological phenomena; in fact, the spell would dissipate only very gradually.

The important contributions of chemists to the gathering of new, precise biological observations were a factor in this retreat: the study of fermentation, the study of the metabolic fate of certain products administered to animals, and the isolation and determination of the structure of molecules produced by living organisms. Thus, the branch of chemistry known as "organic chemistry" was born in the last century. The problem of the boundary between this new chemistry and biology arose immediately as a practical matter. Justus von Liebig (1803–1873), who believed in the existence of the "vital force," devoted a great deal of his time to thinking about biological processes, although he remained a chemist at heart for his entire life. Louis Pasteur (1822–1895), on the other hand, a talented chemist to begin with, became involved in typically biological problems. This was a time when it was possible to move from one discipline to another without enormous institutional difficulties.

Nevertheless, biology and chemistry remained quite separate disciplines for several decades, and at one time it even seemed that the gap would widen still further. The theories of evolution and the observations of heredity that were developed during this period captured the attention of the scientific world; at first glance, neither seemed to have anything at all to do with chemistry. Chemistry, for its part, developed by discovering new "chemical functions," namely new groups of atoms with characteristic chemical reactivity, and new methods of synthesis. It established the basis for the kinetics of reactions so that their rate and their mechanisms could be studied; thus, "stereochemistry" was born, supplying information such as on the spatial arrangement of atoms in a molecule. The dichotomy between the preoccupations, ways of thinking, and experimental techniques of the two scientific fields appeared to be total.

But they were to converge once again. The development of the biological sciences and the overall (macroscopic) observation of events affecting living matter led to a more detailed and more analytical study of these processes with the hope of gaining a better grasp of their intimate mechanisms. "Biochemistry" set itself the goal of accounting for biological phenomena on the molecular level by discovering their chemical bases. First, beginning with whole organisms, the chemical transformations operating in the world of living organisms were studied: a given substance would be administered to an animal and its excretions examined to discover what effect metabolism had on the compound. It was this type of experimentation that allowed Claude Bernard (1813–1878) to define the role of sugar in the animal or human organism, to discover the glycogenic function of the liver (1848), and to study the digestion and absorption of fats (1856).

But further steps were needed before the phenomena studied could be delimited in greater detail: perfusion of isolated organs, then experiments on tissue slices or ground tissue, and finally isolation and purification of biological macromolecules, thus removed from their cellular environments. These methods of investigation appeared to be more and more destructive of biological material but at the same time more and more finely analytical.

With the isolation and purification of biological macromolecules, the ultimate stage—already approximating classical chemistry—was reached: *in vitro* research. The study of enzymes, the catalysts that trigger chemical reactions in living things, made great strides: in many cases it became possible to understand the mechanisms of their action in detail, to the point where they could be represented by reaction schemes similar to those used by chemists.

In general, since the beginning of the 20th century, the study of proteins has gradually tended to take an approach increasingly typical of chemistry. These large molecules, composed of chains of many different amino acids, are the most abundant in living cells and constitute over 50% of the cell's dry weight; it is estimated that there are anywhere from 10^{10} to 10^{12} varieties of proteins in the 1.5 million species of organisms alive today. An exact formula has been established for some of them using modern methods to identify the nature and sequence of the amino acids of which they are composed.

Despite the fragility of these macromolecules, they can be treated like any other chemical compound, using methods such as crystallization, x-ray diffraction, nuclear magnetic resonance, binding to insoluble polymers, and industrial use as catalysts, which were formerly reserved for organic chemistry alone. None of this research has revealed any contradiction of known physical or chemical laws; all the chemical transformations observed can easily be explained by classical laws, and a chemist can reproduce them in the laboratory.

A similar convergence between biology and chemistry or physical chemistry occurred in the study of nucleic acids, another type of basic biological macromolecule. These are "nucleotide" polymers, each nucleotide having a molecular fragment of a basic nature (nucleobase) bound to a sugar, itself attached to a phosphate group. Great progress has been made in the investigation of these macromolecules, particularly since the 1950s, through the use of physical chemistry: working out the structures of deoxyribonucleic acid (DNA), identified as far back as 1869 by Johann Friedrich Miescher (1844–1895); discovery of the mechanisms of polynucleotide biosynthesis; and the organization of chromatin, a multimolecular aggregate of DNA and several proteins.

The role of nucleic acids as carriers of genetic information was demonstrated by O. T. Avery, C. M. MacLeod, and M. McCarty in 1944. The structure of the letters (nucleobases) of the genetic code had been known for a long time, and it was not long (around 1965) before the words or "codons" were deciphered; these are groups of three successive nucleotides that specify the amino acids that comprise the protein to be synthesized. New methods of determining the sequence of bases in polynucleotides made it easier to read the information. Conversely, for "writing" the information, genetics provided a great impetus to chemistry to develop new methods of synthesizing polynucleotide chains.

The gap between biology and chemistry continues to narrow even today. The reshuffling of the biological disciplines and the trend toward greater exactitude in the study of structural and functional aspects of biological processes have given rise to what for several decades has been termed "molecular biology." Progress in this direction has already been spectacular. Precise information is becoming available about the elementary molecular interactions that form the basis for certain biological processes; it is becoming possible to look at them through the eyes of a chemist.

CHEMICAL REACTIONS

In inorganic chemistry, when, for example, a silver nitrate solution is mixed with a potassium hydroxide solution, a double decomposition reaction is triggered, and silver hydroxide is precipitated out. The silver hydroxide forms instantaneously and completely. Indeed, in inorganic chemistry, reactions in solution occur between ions with opposite charges and having spherical symmetry. Hence all collisions between reacting entities are effective, at whatever angle they collide.

This is not the case in organic chemistry. For collisions between the molecules of reacting substances to be effective, two conditions must prevail: the impact must occur with sufficient energy (greater than a minimum value known as activation energy) and the molecules must be properly oriented at the time of the collision.

Because the great majority of collisions fail to meet these conditions, they are ineffective and do not lead to a reaction. As a result, reactions in organic chemistry are generally slow and incomplete, and their yield, namely the ratio between the quantity of product actually obtained and the quantity theoretically expected, rarely reaches one hundred percent.

To speed up these reactions, either of the two conditions stated above can be manipulated. Raising the temperature increases the percentage of molecules with sufficient energy to react; thus, a temperature rise of 10°C (50°F) doubles the reaction rate on the average. This method is used frequently in the laboratory, but it is rarely employed by Nature. When we look at the evolution of living organisms, we see that the appearance of warm-blooded animals (mammals and birds) some 150 million years ago was undeniably a great leap forward. We are familiar with the habit of certain poikilothermic animals (like lizards, whose body temperature varies in response to the environment) of basking in the sun. But although certain species have demonstrated their ability to survive in sometimes astonishingly hot environments, the internal temperature of living organisms does not generally exceed a hundred degrees or so; the maximum is 43.5°C (110.3°F) in certain birds.

Instead, Nature takes the second approach: suitable orientation of the reacting molecules. Chemical reactions proceed far more quickly when the two reacting functional groups are part of the same molecule. This decreases the freedom of movement of these groups while increasing the frequency of collision. If, in addition, the directions of the bonds are properly oriented, the proportion of effective collisions also increases. Also, the activation energy of the reaction decreases and the reaction rate increases: *intra*molecular reactions occur more rapidly than *inter*molecular reactions of the same type. It is possible to calculate what the concentration of molecules in an intermolecular reaction would have to be for its speed to be the same as that of its intramolecular equivalent: this often reaches enormous values, corresponding to highly theoretical concentrations that would be impossible to achieve in practice.

It also appears that, in the case of intermolecular reactions, any physical or chemical interaction that is able to bring the molecules together in the first place favors the reaction by conferring upon it the characteristics of an intramolecular reaction.

In enzymatic reactions, substrate S of the reaction initially becomes attached to the enzyme E to form a noncovalent complex ES ("noncovalent" meaning without a true chemical bond); the reaction that follows this preliminary step is thus always of the intramolecular type. This pre-attachment, due to the complementary structures of the enzyme and its substrate, often involves the formation of "hydrogen bonds" between these two molecules, that is, they share a certain number of hydrogen atoms. Such bonds are weak but precisely oriented to a specific atom in the vicinity; thus they ensure that the enzyme specifically recognizes its substrate. Termed "directional" bonds, there are often so many that together they can contribute to creating a solid attachment and a great affinity of the enzyme for its substrate. Other factors, such as electrostatic attraction between molecular fragments of opposite charges, and hydrophobic interactions attracting uncharged groups, also assist in the formation of the enzyme-substrate complex. Finally, a true covalent bond may sometimes be established between the substrate and the enzyme, analogous to the bonds between the atoms of a molecule; although it is not generally stable, this bonding also facilitates the reaction. The recognition specificity of the substrate may often be "absolute," in which case the enzyme specifically binds to a given substrate, excluding all other molecules. Other enzymes, for example hydrolytic enzymes that catalyze numerous reactions in which organic molecules are split by water, can act on various substrates with a related structure. Even in this case certain signs of specificity are observed, because the rates of these reactions generally vary with the nature of the substrate.

The molecules known as isomers (from the Greek *isos*, "equal," and *meros*, "part") are composed of atoms that are structurally identical and have identical functions but whose properties differ. This is attributed to the difference in the relative arrangement of their atoms or groups of atoms in space. In particular, certain molecules have different optical properties and therefore belong to the class of optical isomers designated by the letter D, or the class designated L. Living organisms are composed of specific optical isomers. Thus, proteins contain only L amino acids while polysaccharides are polymers of D sugars.

Several hypotheses have recently been advanced to explain how the choice of these molecules could have operated in the course of evolution: for example, unequal resistance of isomers to decomposition by beta radiation, crystallization selectivity of isomers, or greater difficulty of polymers in forming when the isomers are different than when the isomers are identical.

Whatever the case may be, the very clear distinction between isomers is one of the fundamental attributes of the chemistry of life. Thus, Nature treats D and L amino acids as if these substances were totally foreign to each other, while the organic chemist tends to view them as related compounds because they are formed by the same synthesis path. Enzyme processes affect only molecules with a given geometric shape, that is, strict stereospecificity is the rule; but this is difficult for the organic chemist to produce by chemical synthesis.

In an enzyme protein, the amino acids responsible for pre-attachment of the substrate and those participating directly in enzyme catalysis are all properly arranged relative to the substrate molecule.

This favorable arrangement can exist prior to the reaction or may be induced by binding of the substrate, causing a change in the geometric shape of the protein, which is a flexible macromolecule. Thus, a number of reactive groups are brought together and properly oriented at the same moment. Such a feat proves impossible in organic chemistry, where the probability of several molecules with favorable orientations coming together at the same moment in time is almost zero! Only enzymes have the ability to organize such complex and chemically effective structures.

But the role of these enzyme molecules does not stop here. We know that the nature of the solvent also has an effect on chemical reactions, depending on whether the substance is polar, hydrophilic, or hydrophobic, and also on such factors as the degree of acidity of the environment (its pH), and the presence of electrically charged particles. As a very brief reminder: chemical processes require energy to temporarily activate the starting materials and this transition state then progresses to formation of the reaction products. The higher the energy level ("free energy") of the transition state, the more difficult the reaction will be, because it will require more activation energy. The free energy of the transition state also depends on the environment in which the chemical transformation occurs. If the transition complex is electrically charged, the reaction will be favored in a polar solvent whose molecules, themselves carrying separate charges, are oriented according to the charge of the complex. This association with the molecules of the solvent (or "solvation") requires work, resulting in a loss of free energy of the transition state and hence a decrease in the free energy of activation of the reaction, which accordingly becomes easier.

Other reactions take place more readily in non-polar or uncharged solvents. This condition is reflected in the structure of the enzyme proteins whose active reaction sites lie in a hydrophilic region or in a hydrophobic pocket, depending on the nature of the reaction that is catalyzed. A hydrophobic pocket creates a microenvironment around the reaction site which plays a role analogous to that of a non-polar solvent in a chemical reaction, whereas the natural macroenvironment of the enzyme is always polar (water).

Likewise, pockets of exceptional acidity or basicity, sometimes very different from the overall pH of the medium, can also form and thus promote acid or basic catalysis. For this purpose, proteins have a range of amino acids available that are either acid or basic in nature; these properties can be fine-tuned even more by their microenvironment.

In addition to this, an appropriate three-dimensional form or "tertiary structure" of the enzyme protein often allows the overall chemical reaction that is catalyzed by the enzyme to be broken down into a series of steps, each of which requires only a small amount of activation energy.

So we see that, in actual fact, it is geometry that is the most important factor in facilitating enzyme reactions; it alone is responsible for increasing the reaction rate to as much as 10^8 times that of noncatalyzed reactions. The contribution of other factors (acid-base catalysis, covalent catalysis, etc.) appears to be more modest, although they are still able to increase the reaction rate a thousandfold.

Finally, it should be remembered that the increase in the rate of reactions catalyzed by enzymes can range from 10^8 to 10^{20} times that of the same reactions without catalysis. Although they are complex, enzymatic reactions can thus become very rapid, like inorganic reactions. There is no infringement of the first law of thermodynamics (conservation of energy) nor of the second (increase in the entropy or, roughly speaking, the disorder of a system). Enzymatic catalysis eliminates the risk of the disorderly and usually ineffective random encounters between molecules which characterize the noncatalyzed organic reactions in inanimate matter.

CHEMICAL FUNCTIONING OF THE CELL

Here we have, in broad outline, the highly evolved system of the chemistry of life. The next question is: what is this system used for?

Its purpose is to decipher and implement the genetic message by creating the very substance of living matter, principally biological macromolecules. A parallel purpose is to obtain the chemical energy necessary for completing these processes. The synthesis of macromolecules occurs in two stages. The first involves production of small molecules, that is, amino acids, nucleic bases, sugars, etc. In the second stage, these "building blocks," as they have come to be called, are assembled into polymers. Heterotrophic organisms, such as animals and most microorganisms, so called because they derive their nourishment from other living organisms by using existing molecules, acquire the energy and raw materials needed by breaking down foods. Autotrophic organisms, such as photosynthetic cells and certain bacteria, manufacture biological macromolecules by synthesis from mineral elements; thus their chemical activity appears to be more complete and more diversified than that of heterotrophic organisms.

The small molecules that must be produced first are not large in number. Nonetheless, each must be synthesized in several stages from simpler substances, and the cell must manufacture a large number of compounds to accomplish this. Consequently, the cell abounds with all manner of molecules: the famous bacterium *Escherichia coli*, a model of simplicity, contains about five hundred kinds of small molecules in addition to three thousand proteins and a thousand or so nucleic acids! At all times, this tiny cell is teeming with precise, perfectly orderly and effective chemical reactions; at the same time, these reactions must be very fast because the lifetime of a cell can sometimes be no longer than twenty minutes. Nearly all of these reactions can be reproduced in the laboratory because they obey the laws of chemistry and thermodynamics in every respect.

But how can such a large number of chemical processes occur simultaneously in such a restricted space, while a chemist can carry out only one reaction at a time using a single vessel as the reactor? This extraordinary performance is made possible by the specificity of enzyme reactions. Each enzyme actually has two extreme and opposite properties at the same time: enormous reactivity to its substrate, but at the same time near-total inertia with respect to all other molecules. Because of this indifference of enzyme reactions to surrounding chemical events, everything actually proceeds as if each of the numerous reactions were taking place alone within the cell according to the principle of "one reaction, one reactor."

These individual "component parts" thus manufactured in the cell then assemble into macromolecules. The stage of fine, diversified chemistry in which small molecules are created is followed by a long series of repetitive reactions of a uniform chemical nature whose general goal is to join two molecules together by extracting one molecule of water.

This apparent simplicity conceals mechanisms that are actually more complex. According to the general strategy of the chemistry of life, molecules align themselves properly before uniting. The monomers that will be incorporated one at a time into the growing macromolecular chain are sorted with the greatest precision. Thus, to enter into the composition of an incipient protein, amino acids first attach to small ribonucleic acids called transfer RNA (tRNA); the enzymes which perform this reaction, called aminoacyl-tRNA synthetases, operate with enormous specificity and a tiny margin of error. But even so, chemical monotony appears: a whole battery of these enzymes carries out the same reaction, that of attaching a given amino acid to the corresponding tRNA.

This monotony is extraordinary. The process of protein biosynthesis takes place according to the same scheme throughout both the animal and the plant kingdoms! The information about the structure of the proteins that a living organism must synthesize is written in its genome—a central database that constitutes an enormous library in which the genes are the individual volumes; many of them will contain the programming for the structure of a single protein.

The language used by this information storage system is usually that of DNA. The significant elements, four bases called A, T, G, and C (adenine, thymine, guanine, and cytosine), are grouped into codons and attached to a long chain in which a sugar (deoxyribose) alternates with a phosphate. No volume ever leaves the central library; if it is needed, a copy of the DNA is made in the form of RNA with ribose replacing the deoxyribose and the U base (uridine) replacing the T. This RNA, the messenger of genetic information, will undergo a few transformations; then, after this maturation, will be translated into a protein by the ribosomes, complex cytoplasmic particles composed of several tens of proteins and a few RNAs.

The total process thus represents the *translation* of a sequence of nucleotides in a nucleic acid into a sequence of amino acids in a protein; it includes one bilingual "hinge molecule," tRNA, which has both a binding site for an amino acid and a triplet of bases (anticodon) complementing* the codon of the mes-

* The A and G bases are complements of T (or U) and C, respectively; thus, in a nucleic acid double helix, the base C of one strand is linked by hydrogen bonds to G in the opposite strand and A is linked to T (U).

senger RNA (mRNA) that determines the location of the amino acid in the protein being formed.

In the biosynthesis of nucleic acids, the mechanisms are slightly less complex, at least in principle. Processes such as DNA replication or DNA ⇄ RNA transcription do not require translation: they take place within the family of nucleic acids and consist in forming a sequence of nucleotides complementing a strand that models a nucleic acid. An analogy is photographic reproduction, which copies an image onto a negative, which can produce a "positive" image, which can in turn produce a negative image, and so forth.

To the two stages of cellular chemistry described thus far—synthesis of small molecules and polymerization of these molecules—another process is sometimes added: the assembly of macromolecules into supramolecular structures (oligomeric proteins composed of several chains of amino acids, ribosomes, etc.). Such self-assembly is brought about mainly by hydrophobic interactions, which are important but somewhat unspecific because they are nondirectional. They simply result from the tendency of water molecules, when in contact with macromolecules (water of solvation), to dissociate from them, which results in an increase of entropy. Hydrogen bonds and ionic interactions, which are directional and thus ensure good geometric specificity of the assembly so formed, also contribute to producing this third stage in the elaboration of macromolecular systems.

Although this stage cannot be termed "chemical" in a strict sense, it does have a chemical basis because the one-dimensional information of DNA incorporates the information on all macromolecular structures. These include the structure known as "quaternary," namely the spatial arrangement of macromolecules forming supramolecular assemblies.

One can scarcely fail to realize that this synthesis strategy resembles that adopted by the chemist in a laboratory. The chemist too begins by establishing a stock of raw materials in the form of small molecules obtained by extraction from a natural substance or by way of synthesis; these small fractions are then assembled into larger molecules. The same applies to very lengthy syntheses. Thus, for example, chemists do not synthesize long polypeptide molecules all at once: they prepare several short chains, then join them together. The biosynthesis of proteins, a far more sophisticated process, can produce far longer polypeptides by joining amino acids together one by one, but even in this case there is a limit to the length of the chains that can be formed. We note that proteins of high molecular weight are usually oligomeric, that is, composed of several subunits synthesized individually and then joined together. Through this unique strategy, Nature, just like the chemist, can do a great deal to protect itself against errors in synthesis and, if necessary, eliminate malformed fragments.

And so we have a snapshot of the "chemistry of life" as we know it today. But we can look back, behind this snapshot, at its evolution. The highly complex systems of today's living matter grew a step at a time. At each step, the transition to a higher degree of organization was always the recombination of a small number of simpler chemical entities.

II

THE ORIGIN OF LIFE

What a wondrous spectacle is the complex and highly evolved chemistry of the living world! How could such extraordinarily sophisticated systems be put into place? Reflections on the relationship between chemistry and biology are today leading researchers to think with renewed interest and novel arguments about the age-old question of the origin of life. When the possible pathways taken by the slow evolution of matter from the inanimate to the living are explored, numerous phenomena situated at the interface between chemistry and biology are discovered. The once distinct boundary between processes deemed to be "chemical" and those qualified as "biological" seems to be blurring before our very eyes. Does not the knowledge we now have about the origin of life lead us to challenge the distinction between a purely chemical period of evolution followed, after the appearance of the earliest "living" structures, by biological evolution? Would it not be better to imagine a single period of "chemical evolution" including that of living organisms?

Since the earliest days of the Universe, matter has been organizing and evolving toward increasingly complex forms. But what drives this evolution? Today, everything seems to indicate that it was driven from the very outset by natural selection due to environmental pressures—in short, that the Darwinian scheme can be extended to processes unknown to Darwin himself. This selection is in no way teleological (it does not assume a purpose); it naturally eliminates the least stable structures, those least adapted to a given environment, and this process of elimination ensures more or less regular improvement of the organization of matter. It must be recognized that such phenomena become apparent even before we begin studying the origin of life when we wonder about the origin of the earliest molecules—those formed at a time which could be called the "protochemical" period of the Universe by analogy with the proto- or prebiotic period.

THE SMALL PRIMORDIAL MOLECULES

The origin of the Universe witnessed the nucleosynthesis of two primordial elements: hydrogen and helium. The other, heavier elements appeared only later by nuclear fusion in stars. About ninety elements had sufficient stability to ensure their existence, but only a small number of them could arrive at a higher level of organization by generating polyatomic molecules. This is the case with hydrogen, oxygen, nitrogen, and carbon (symbolized, respectively, by the letters H, O, N, and C), as well as a few other elements which have the property of easily forming covalent bonds by pooling electrons. The elements H, O, N, and C require 1, 2, 3, and 4 electrons, respectively, to complete their outer electron shell and form covalent bonds; the lighter these elements are, the more stable these bonds will be. But atoms such as C, N, O, and S (sulfur) can share more than one electron doublet. Double or even triple bonds can form in this way. Carbon occupies a very special position among these atoms: it can complete its outer electron shell by donating or, conversely, accepting 4 electrons; thus it is capable of forming covalent bonds with 4 other carbon atoms and making long straight or branched chains, as well as rings, either single or fused with other rings. As a result, a huge number of molecules may be formed, with a wide variety of three-dimensional structures. To this we should add that carbon is also capable of forming similar single or multiple covalent bonds with oxygen, nitrogen, sulfur, and other elements: hence the numerous different types of chemical functions.

So it comes as no surprise that the majority of polyatomic chemical species found in interstellar space are composed of carbon, hydrogen, nitrogen, and oxygen.

A more detailed examination of these species reveals the influence of the environment on the type of polyatomic assemblies thus formed. The Milky Way remains the best-studied area of interstellar space. Astrophysicists have taught us that ninety percent of the mass of our Galaxy is concentrated in its 10^{11} stars; the remaining ten percent is accounted for by gases, mostly hydrogen, and dust, with an average density of a few particles per cubic centimeter. Molecules have been detected inside these gases in the densest regions called "interstellar clouds," which are diffuse, dark, or black.

In the diffuse and relatively hot clouds, highly intense cosmic ultraviolet radiation causes rapid photodissociation of the molecules, which are consequently rare and reduced to a few diatomic species.

The dark, denser, and cooler clouds, which are enclosed within the diffuse clouds, are composed of gases and cosmic particles that partially block stellar ultraviolet rays, the agents of molecular dissociation. Hence, the dark clouds may be the site of reactions and consequently of some degree of chemical evolution. Stable molecules such as H_2, CO, H_2CO, HCN, and NH_3 are found in addition to unstable polyatomic species (HCO^+, HN_2^+, HC_2, HNC, C_3N, and C_4H). However, the presence of residual ionizing radiation (ultraviolet, x-rays, or cosmic rays) in these clouds prevents the formation of the molecules one might expect to arise from thermodynamic equilibria in a hydrogen-rich environment, namely methane (CH_4) and water (H_2O).

But these dark clouds contain numerous black clouds. At the center, the forces of gravitation and higher temperature lead to the gestation of a new star. The nature of the environment (high temperature and density, low radiation) favors a more advanced type of chemistry and the production of more complex molecules such as ethyl alcohol (C_2H_5OH) or ether (CH_3-O-CH_3) which are stable under these conditions.

Now, in many of the organic interstellar molecules we find compounds which are precursors of biological molecules: hydrogen cyanide, which can generate amino acids and nucleic bases; formaldehyde, the precursor of sugars; cyanoacetylene, an important condensation agent; etc. These molecules are able to form even under extreme conditions of temperature and high concentration of interstellar media. Apparently ubiquitous in the Universe, they must certainly have existed on the surface of the primitive Earth, as well as on other planets: traces of amino acids, which are already more complex chemicals, have been identified in lunar dust and in meteorites. If to this fact we add purely statistical arguments having to do with the immense number of celestial bodies that exist, it seems highly improbable that the Earth is, or was, the only place in the Universe where life developed, even though most studies devoted to the search for the origins of life have thus far dealt with its appearance on our own planet. Here we should remember that the panspermia hypothesis of Arrhenius, according to which life appeared on Earth because of spores carried from other regions of the Universe, does nothing to account for the origin of life in general; it merely shifts the problem. At the present time, what is known as "exobiology," namely the search for possible life forms in the Universe, is an important field of activity at NASA. A research program entitled SETI (Search for Extraterrestrial Intelligence) has brought together the efforts of the former Soviets with those of the Americans to develop a powerful detector for this purpose.

This brief look at the first phases of organization of matter has brought us as far as the precursors of biological molecules; it shows that natural selection, under environmental pressure, was already operating at a very early stage of evolution.

PRIMITIVE BIOMOLECULES

As we have seen, only a small number of elements which are capable of joining by covalent bonds and thus giving rise to an infinite variety of molecules could lead to a higher degree of organization of matter. The other elements could only enter into a more impoverished type of chemistry, mainly based on ionic or coordination interactions. We can see that this dividing line between the few elements with a high evolutionary potential and all the others continues to separate the living from the inanimate. Living organisms are mainly composed of the elements H, O, C, N, P (phosphorus), and S. A few inorganic elements also prove to be indispensable to life but they are involved only in the form of ions (namely atoms that are electrically charged by the addition or subtraction of one or more electrons) in accordance with their chemical nature: these are calcium, chlorine, potassium, sodium, and magnesium. Finally, sixteen other elements (manganese, iron, zinc, copper, cobalt, selenium, molybdenum, iodine, vanadium, silicon, tin, boron, aluminum, nickel, chromium, and fluorine) may play a role in living organisms but they are not all essential to every single species.

So, although the appearance of life brought about a powerful shift of matter, only a small number of elements was involved. Hence, the striking differences in composition between the lithosphere (the upper part of the Earth's mantle) and the biosphere: while hydrogen (93.4%) and helium (6.5%) predominate in the Universe in general, the Earth's crust contains mainly oxygen (33%), iron (27%), silicon (17%), magnesium (14%), sulfur (2%), nitrogen (1.6%), calcium (1.3%), aluminum (1.2%), and a few other metals as well as a maximum of 0.22% hydrogen. In living organisms, the proportions change once more: the human body is composed of 63% hydrogen, 25.5% oxygen, 9.5% carbon, and 1.4% nitrogen; the other elements are present only in far smaller quantities.

Since the primordial molecules contained only a small number of atoms, they necessarily had to be joined by multiple bonds to meet the valence requirements. Thus, for example, in the hydrocyanic acid molecule, H–C≡N, composed of three atoms, carbon and nitrogen require a triple bond, while in the penta-atomic cyanoacetylene molecule, N≡C–C≡CH, two triple bonds are required. Multiple bonds break with relative ease: they are severed by the addition of molecular fragments with entire or partial charges of opposite signs to both sides of the multiple bond, as for example those produced by the dissociation of an acid AH into a proton H^+ and an anion A^-. Some of the primordial dissociable molecules (HCN, H_2O, etc.) could be added to the multiple bonds; the HCN molecule, which is both dissociable and unsaturated, is able to play the role of "donor" or "acceptor."

The first molecular structures, most of them small, harbored an enormous evolutionary potential, as we have seen, so that additions to the multiple bonds may have had a truly biogenic action. The transformation of hydrocyanic acid into cyanoacetylene and cyanamide, then into amino acids and nucleobases, proceeds from this reaction, like the transformation of formaldehyde into sugars. These processes are not simply hypothetical; they have been experimentally verified on numerous occasions under conditions simulating those of the primitive Earth. This was in particular the result of the famous 1953 experiment by Stanley L. Miller in San Diego on the formation of amino acids.

By combining with each other, these small molecules made considerable inroads into their reactivity reserves, but an important step in the chemical evolution leading to life had been taken, since these reactions generated compounds that were the building blocks of life.

We should add that only a small number of such building blocks are needed to form living structures. The thirty biomolecules that can be classified as truly fundamental are: twenty amino acids, five nucleobases, two sugars, glycerol, an amino alcohol (choline), and a fatty acid (palmitic acid). Of course, the number of small molecules produced by living cells that exist today is far larger. Nonetheless, the 150 natural amino acids, the few dozen sugars and nucleotides, and the numerous fatty acids that have been found in living organisms today are all derived from this group of thirty or so fundamental biomolecules! In the same way, most alkaloids have been formed from amino acids; rubber, terpenes, vitamins, essential oils, and plant pigments all come from glucose or fatty acids through the intermediary of acetic acid. These molecules certainly made their appearance in the later stages of evolution. Most of them result from complex biosynthetic processes, involving the concerted action of numerous enzymes, that proceeded according to the rules of systems that were already highly elaborate.

The history of the Earth definitely witnessed an epoch when, side by side with the original lithosphere, large quantities of organic molecules were able to accumulate, constituting what could be called an "organosphere." It is difficult to evaluate the concentration of organic compounds in the primitive ocean because their rates of synthesis and decomposition are still poorly known. Nonetheless, an approximate calculation worked out for amino acids shows that the concentration of these compounds, formed with an estimated ten percent yield from hydrocyanic acid, must have reached 3×10^{-4} gram molecules per liter after ten million years. Of course, this is an average value; it is generally accepted that the concentration of organic substances in the primitive ocean varied from region to region, but local conditions could cause these compounds to be present in large quantities at certain sites, facilitating their interaction. In this way, the organosphere was able to evolve toward a higher degree of organization and gradually give way to a biosphere in which organic matter was no longer distributed randomly but grouped to form living organisms. As time went on, the biosphere in its turn was able to engender a new type of organosphere which, by contrast to the first, came about not from chemical evolution toward greater complexity but from the products of degradation and disorganization of the matter of which living organisms are composed, namely coal and petroleum.

THE MACROMOLECULES

These, as we see them today, were the two stages that punctuated the formation of the earliest basic molecules of life: the birth of tiny molecules which were able to form even under conditions where matter was very thinly spread, as in interstellar space; and mutual interaction of these molecules leading to more complex structures, stable under the condensed matter conditions of the primitive Earth. Abiotic formation of these building blocks of life abided by the laws of thermodynamics. But it only supplied the basic elements—the alphabet of life, as it were. What might be the chemical future of these elements?

For evolution to continue, a new biogenic reaction was then required. The structure of these early biological molecules actually allowed them to undergo one more reaction: that of dehydrating condensation which had the effect of joining them together. However, such a reaction was impossible in the primeval oceans because it cannot occur spontaneously in the presence of water. A condensation agent was needed to extract the elements of water from the two reacting molecules. It turns out that certain nitrogenous molecules can act as condensation agents (cyanogen, cyanoacetylene, etc.). And we now know that they were present on the primitive Earth, because they have also since been found in interstellar space. Moreover, the addition reaction on the multiple bonds that led to the earliest organic molecules may also have engendered several condensation agents: in this way, hydrocyanic acid would have yielded cyanoguanidine and amidinium carbodiimide. The experiments mentioned above that simulated the conditions prevailing on the primitive Earth showed that a condensation agent such as cyanovinyl phosphate is formed by cyanoacetylene joining with phosphate. These experiments also indicated that nitrogenous dehydration reagents quite possibly induced the formation, under gentle conditions, of polyphosphates and polyphosphoric esters which appeared as the most important biological condensation agents, the precursors of adenosine triphosphate (ATP).

Condensation agents played a dual role in the chemical evolution that led to the appearance of life. First, by eliminating one molecule of water between two existing organic compounds, such as the purine or pyrimidine bases, sugars, phosphates, fatty acids, and alcohols, they created new and more complex categories of fundamental biomolecules of living matter, in particular nucleosides which result from the attachment of a sugar to a nucleic base, nucleotides which are phosphorylated nucleosides, phospholipids, etc. More important still, by combining several building blocks in the same way, condensation agents made possible the formation of an infinite variety of prebiotic macromolecules.

The issue of the chronological order in which the two important types of macromolecules—proteins and nucleic acids—were formed has given rise to lively discussions over the years. The "proteins first" proponents point out that numerous experiments, particularly those performed in the 1950s by the American Sidney W. Fox, showed that amino acids were capable of polymerizing fairly easily to yield polypeptides under various conditions simulating those which may have prevailed on the primitive Earth. This polymerization may have been induced by electrical discharges, by heat (geothermal energy, for example), or by contact with certain types of clay and polyphosphates. Such primitive polypeptides, which Fox called "proteinoids," may reach quite a respectable size and contain up to 150 to 200 amino acids. These spontaneous polymerizations do not occur entirely by chance; a certain selectivity is observed in the condensations, and the composition of the peptide chains formed often differs considerably from that of the initial amino acid mixture.

The weak point in this hypothesis which accords chronological priority to the proteins is that proteins are not informational macromolecules and therefore cannot multiply by themselves. So what could have been the mechanism by which their structure was memorized? Conscious of the need to answer this crucial question, Fox advanced a number of tentative explanations. The fact that amino acids do not polymerize at random to produce proteinoids already indicates some selectivity, he noted, and constitutes the first step in the evolution of genetic information. The next step, Fox postulated, would be the formation of aggregates of peptides with a nonrandom structure, which he called "microspheres." Moreover, he showed that proteinoids rich in certain amino acids combine selectively with a given type of "homopolynucleotides," nucleic acids so called because all their bases are identical in this case. So, the rudimentary chemical information contained in the proteinoids may serve to select these polynucleotides and it could have been gradually transferred to nucleic acids. The specificity of the interaction between polypeptides and polynucleotides could have been the priming factor of the genetic code.

The opposite hypothesis, giving chronological priority to the nucleic acids, also had its adherents. Indeed, how could one fail to note the fundamental role played by nucleotides in most biological processes? These molecules are to be found everywhere in living organisms today. They exist in a polymerized state in DNA and in the three types of RNA. Nucleotides play a primordial role in the chemistry of life, where they ensure the transfer of energy, hydrogen or electrons, sugars or lipids. We can see that they are also involved in all the essential domains of life: the genetic apparatus, metabolism, and energy transfer. Moreover, the role of nucleic acids, even in the living world of today, is not exclusively confined to that of informational molecules; two types of ribonucleic acids, the ribosomal RNAs (rRNAs) and the transfer RNAs (tRNAs) play a functional role. In the biosynthesis of proteins, the participation of ribonucleic acids is still seen as predominant today; they account for two-thirds of the composition of ribosomes in bacteria. The fact that numerous coenzymes, small molecules that assist in enzymatic catalysis, are nucleotides or nucleotide derivatives has led some researchers to regard these coenzymes as the fossils of the most ancient polynucleotide enzymes.

However, the objection that can be raised to the "nucleic acids first" hypothesis has to do with the difficulty of forming nucleotide constituents in the primeval soup.

We will not list all the arguments of those who, like Fox, are proponents of the "proteins first" hypothesis or those who, like Nobel Prize winner Francis H. C. Crick and Leslie F. Orgel believe in the "nucleic acids first" hypothesis. Both scenarios are based on equally sound and plausible reasoning; neither, however, convincingly solved the fundamental contradiction which poses the classic "chicken and egg" problem: DNA is needed to store information on the structure of proteins, but proteins are needed to perform DNA synthesis. Biologists, who reason from observation of today's living world, divide up the main macromolecules roughly into two distinct categories: informational molecules and functional molecules. So it is hardly surprising that this doctrine of separate roles for nucleic acids and proteins seemed to many people to go without saying. The logic of the system now observable is perfectly clear. *A priori*, it appeared difficult to attribute an important functional role to a chain of nucleobases that had little variability or reactivity, attached to a more or less inert sugar-phosphate chain in the manner of notes written on a musical staff. On the other hand, the absolutely essential meaning of the *sequence* of these bases was obvious. I venture to employ this musical metaphor which actually owes nothing to chance: the transcription of gene sequences in music where a given note corresponds to a given nucleic base and the psychosomatic effect of this "DNA music" have been the subject of research by a few pioneering spirits for several years.

There appeared to be an equal logic in the catalytic activity of proteins which are composed of numerous types of amino acids. Proteins, operating alone or in concert with other residues in the vicinity, constitute chemical tools of unmatched fineness. Others, having no chemical function in their side chains, contribute to maintaining the spatial structure of the entire molecule and ensuring its effectiveness.

A number of attempts have been made to break this vicious circle. In 1968, Crick came up with the idea that the earliest ribosomes could have been composed entirely of RNA. The first "enzyme" would have been made of an RNA molecule endowed with replication properties. The famous Nobel Prize winner believed it possible that Nature first had RNA play the role later taken over by proteins. At the time, this supposition was only a theoretical construct with no experimental backing, and the problem remained unsolved.

During the last few years, however, a solution seems to be emerging. We may finally be getting to the bottom of the puzzle. Some recent papers, in particular those which earned Thomas R. Cech and Sidney Altman the Nobel Prize in Chemistry in 1989, actually demonstrated that informational and catalytic properties could really be combined in the same molecule, a polynucleotide. So we now have good reason to assume that the oldest polymers were the RNAs which in the beginning would have played the dual role of repositories of genetic memory and catalysts. Viewed from this perspective, the egg *is* the chicken.

The early RNAs presumably favored the development of proteins, which then confidently stepped into the role of biological catalysts and—thanks to reverse transcriptases (enzymes that synthesize DNAs with a sequence of bases equivalent to that of a model RNA)—presumably transferred the genetic information to DNA, a polynucleotide more stable than RNA. The functions performed by RNAs today, important though they are, particularly in protein biosynthesis, are a pale reflection of the primordial and universal role that these macromolecules once played.

It must be said, however, that these new theories on the primacy of RNAs in evolution leave chemists perplexed. While experiments simulating chemical events under the conditions of the primitive Earth demonstrate the possibility of spontaneous formation of nucleobases, amino acids, and polypeptides, the appearance of RNAs is more difficult to explain. How are we to account for the formation of ribose, not to mention deoxyribose? A detailed study of the formation of sugars from formaldehyde shows that this reaction actually leads to about fifty different carbohydrates, among which ribose ranks only as a minor substance. So today's chemists are imagining, and testing experimentally, the possibility that the backbone of the primitive nucleic acids was not initially a ribose-phosphate chain but was composed of molecules that formed more easily. We know that polynucleotides with ribose-phosphate backbones are not the only compounds able to yield double helixes by base pairing. We know of nucleotide analogs that can undergo polymerization catalyzed by ordinary polynucleotides, and result in nucleotide chains. Although they have an unusual backbone, such chains nonetheless prove capable of hybridizing with classical polynucleotides having a complementary sequence. This is the case, for example, of certain nucleotides obtained by chemical synthesis in which glycerol replaces ribose; it is therefore not impossible that the first nucleic acids may have had such a simplified structure.

Whatever the history and mechanistic details of the formation of these biopolymers may actually have been, we can see that a fairly modest and limited chemical evolution may have been sufficient to lay down the foundations for life. Simple chemical laws account for the transformation of tiny primordial molecules into more complex structures composed of several dozen atoms: the fundamental biomolecules and condensation agents. The interaction of these two types of compounds leading to macromolecules opened the way for a very important step to be taken in evolution. A small number of building blocks could have generated an immense number of polymers; the extraordinary complexity of the living world results from the combination of this small number of basic elements. (At this point, we cannot refrain from referring to another illustration of the infinite richness derived from a combination of a small number of basic elements: language. Every idea can be expressed with the twenty-six letters of the alphabet or even with two signs, as in the Morse code and the binary system used in computers.) Indeed, the formation of heteropolymers introduced an unparalleled richness of molecular structures and properties. The two main classes of biopolymers are constructed on the same model: a regular, repeating backbone with variable side chains, consisting of four types in the case of polynucleotides and twenty types in the case of polypeptides. This type of structure may generate an astounding number of polymers with ν monomers: 4^{ν} polynucleotides and 20^{ν} polypeptides. Thus, for a polynucleotide with 300 bases, coding for a small protein, the number of theoretically possible sequences is 4×10^{180}, which is more than the total number of electrons in the Universe! The value of v is 3×10^3 to 5×10^5 for viruses, 10^6 to 10^7 for bacteria, and 10^9 to 10^{10} for plants and animals. In actual fact the total number of proteins that exist presently in all living organisms does not exceed 10^{12}, since these proteins have common ancestors. By comparison with these gigantic, literally "hyperastronomical" magnitudes, the large numbers found in chemistry seem almost small: the famous Avogadro's number establishes the number of molecules contained in 1 gram molecule of a chemical compound as 6×10^{23}; the number of all known organic compounds does not exceed 10^7.

SUPRAMOLECULAR STRUCTURES

Let us take one more step in the reconstruction we have begun. The prebiotic macromolecules gradually formed into thermodynamically stabler supramolecular assemblies. We know some of the processes which may have presided over these rearrangements.

They may have derived from hydrophobic interactions of the type discussed above; in these interactions, nonpolar groups join to form micellar structures, that is, molecular clusters containing these nonpolar groups at the center, while the polar parts point outward toward the aqueous environment. This phenomenon is analogous to the spontaneous formation of membranes and micelles, which is characteristic of fatty acids and lipids. We know that aqueous solutions of hydrated polymers can generate separate droplets. Fox *et al.* observed that proteinoids, obtained by thermal condensation of amino acids, agglomerated into "microspheres" two microns in diameter when the solution in which they formed was cooled, at least if the pH and the salt concentration were appropriate. The outer surface of such a microsphere has a double-layered structure formed by nonpolar groups of amino acids; it constitutes the analog of a lipid barrier resembling a natural membrane. The behavior of microspheres is in some ways reminiscent of that of living cells; indeed it has been observed that, still under given pH and salt concentration conditions, they may be made to divide. They may also undergo a budding process: one can see a bud gradually detach from the original microsphere and end up as an independent droplet.

Aleksandr I. Oparin, since the 1920s one of the pioneers in biochemical research into the origin of life, reported a phenomenon called "coacervation," by which droplets containing macromolecules are formed. What is happening is spontaneous separation of an aqueous solution of a polymer into two phases, one polymer-rich and the other polymer-poor. In all likelihood, coacervation, followed by emulsification into droplets or "coacervates" of the high-concentration phase (containing five to fifty percent polymer), could have occurred in what he called the Earth's "primeval soup." These droplets, which are closed systems, possessed thermodynamic stability properties which varied according to the environment in which they were located. Only the most stable, having minimum free energy and maximum entropy, were able to survive; the others dissipated into the solution. Thus, environmental pressure also continued to play its selection role at the stage in which these droplets of concentrated material were formed. In general, each step forward in chemical evolution would correspond to a gain in entropy by the system and its aqueous environment. The tendency for the entropy of the Universe to increase has always driven this evolution; it has led to entities that were fitter and more likely to survive. So at this stage, as at all others, we find the phenomenon of adaptation to the environment: although it is a characteristic of the living world, it was also manifested throughout prebiotic chemical evolution.

SIMPLE FUNCTIONS

The progressive evolution of matter toward molecular forms of increasing complexity gave rise to a parallel development of new functions. The molecules that appeared successively in the various stages of organization were not inert in terms of chemistry or physical chemistry; if they had been, evolution would have come to a stop. So this reactivity allowed them to participate in the building of new molecular species and to interact with each other. This reactivity had therefore both a structural and a catalytic role, and it diversified as the size of the molecules increased. As we have seen, even the tiny primordial molecules possessed chemical functions that allowed them to be transformed into more complex compounds which, in turn, were endowed with the ability to polymerize and generate macromolecules. According to the recent research of which we have outlined the main points above, it seems that the polynucleotides were not only the oldest biological polymers but also the first catalysts. Hence, the chemical development of matter could be attributed first to RNAs, whose reactive potential is distinctly less than that of proteins. If we follow this line of thinking, we see the relevance of current ideas on the way in which the reactivity of RNAs, albeit limited, may have contributed to chemical evolution by leading to new, improved structures and functions.

In the last few years, several teams of researchers have discovered RNAs with catalytic properties. The pioneer in this field, Thomas Cech, also showed that in the protozoan *Tetrahymena thermophila*, during the maturation of ribosomal RNA, a fragment of the polynucleotide chain called "intron" or IVS (intervening sequence) of 413 nucleotides was precisely excised from the precursor RNA without, however, the intervention of an enzymatic protein. Newly synthesized RNAs, as we have seen, usually undergo a maturation process which consists mainly of "splicing," or enzymatic excision, of certain fragments of the polymer chain. But in the present case we have "self-splicing" since the process occurs without the intervention of enzymes; only the presence of the nucleotide guanosine triphosphate (GTP) or any guanosine derivative is necessary.

In-depth studies have been made of the sequence and mechanism of this reaction. Self-splicing of *Tetrahymena* pre-RNA occurs through a set of cleavage and ligation reactions which preserve the total number of "phosphodiester" bonds, that is, bonds by which a phosphate group connects two neighboring nucleotides. On the other hand, classical splicing of an intron by an enzymatic protein consists of two cleavages of the phosphodiester groups and one ligation. In addition, this classical reaction cannot ordinarily occur without the intervention of ATP to supply energy, while self-splicing in *Tetrahymena* takes place with no energy supplied from outside.

This aspect of the reaction is accounted for by a process of "transesterification." (The action of acids on alcohol, with water being eliminated, is known as "esterification.") Such transesterifications do not greatly modify bonding energy and take place with no major changes in enthalpy, the thermodynamic parameter related to these energies. The reactions in which the phosphodiester is split and ligation occurs are "concerted": the first supplies the energy necessary for the second. The force which impels these reactions past the state of equilibrium and drives them to the endpoint is thought to come rather from a change in entropy, itself due, for example, to differences in stability between the initial and end products of the reaction, as a result of changes in the tertiary structure of these RNAs.

The excised linear intron then undergoes a series of other auto-catalyzed intramolecular transformations of the same type. At one of its ends there is another split which breaks off a chain of fifteen nucleotides, followed by cyclization of the rest of the intron by formation of a phosphodiester bond between its ends. Thus the molecule assumes a ring shape. At an alkaline pH, this shortened, circular intron (IVS-15) splits open at the point where cyclization (ring formation) occurred and, by a similar mechanism, excises yet another terminal tetramer, while the rest of the polynucleotide (IVS-19) forms a ring and then, because of the alkalinity of the medium, splits at the point of cyclization.

In this series of reactions, the splicing and cyclization sites can be considered intramolecular substrates of the "enzymatic" activity of the IVS. The term "enzymatic" has to be in quotation marks because a catalyst in the strict sense of the word is not itself changed by the reaction it accelerates. The end product of these transformations, linear IVS-19 (or L-19 IVS) no longer has any cyclization sites and can no longer undergo intramolecular reactions, and the cascade of events stops right there. However, this shortened polynucleotide retains its activity and can exercise it *inter*molecularly by catalyzing cleavage and ligation reactions with other RNAs. Here we have true catalysis in the classic sense of the term: the polynucleotide L-19 IVS acts like a true enzyme and becomes known as a "ribozyme."

One of the reactions that the ribozyme can catalyze is deserving of particular attention. This is polymerization of RNAs. Thus, for example, L-19 IVS is capable of converting a chain of five cytidine nucleotides (C_5) into longer polymers, up to C_{30}, by a series of transformations during which a C_5 chain is extended by one nucleotide at the price of another C_5 molecule being shortened: $2\ C_5 \rightarrow C_6 + C_4$. These thermodynamically neutral transesterifications progress through the increase in entropy due to formation of a complex mixture of substances from a single starting material. The specificity of the ribozyme relative to short chains composed only of cytidylic nucleotides is based on the possibility of hybridization between C_n and a complementary IVS sequence or “internal guide sequence” playing the role of the site on the enzyme to which the substrate binds.

So, in actual fact, the L-19 IVS molecule acts similarly to several enzymatic proteins that convert nucleic acids.

Cech also hypothesizes that an RNA with catalytic properties similar to those of the IVS could use an external RNA instead of the internal guide sequence to synthesize a complementary sequence. One of the strands, designated (+), coming from the dissociation of a double-stranded RNA and having RNA-polymerase properties, could thus self-replicate using the complementary (–) strand as a model.

Among the other ribozymes described recently, we should note the case of ribonuclease P. This enzyme, isolated from *Escherichia coli* and from *Bacillus subtilis* by Sidney Altman, is responsible for the maturation of the 5' ends* of the tRNAs by specifically hydrolyzing a phosphodiester bond in a precursor RNA. As in the case of Cech's ribozyme, the scission affects the P-O bond at 3' and produces 5'-phosphate and 3'-OH end

* In a ribose fragment of a nucleotide, the positions of the carbon atoms are numbered 1' to 5'. In a polynucleotide, the phosphodiester bond connects the hydroxyl group OH in the 3' position of the ribose of a nucleotide to the OH at 5' of the ribose of the neighboring nucleotide. Hence, the first residue in the chain has a free OH at 5'and the last one has a free OH at 3'; this provides the terminology for the ends of a nucleic acid.

groups. The enzyme is a nucleoprotein complex in which the role of the enzyme is played by the ribonucleic component called M1 RNA, while the protein subunit constitutes only the cofactor, necessary for catalysis under certain conditions to modify the properties of the catalytic agent by creating an ionic environment favorable to the reaction.

Altman and his colleagues have even come up with a hypothetical mechanism for this hydrolysis, based on a still simple chemistry derived from inorganic chemistry, that of the lithosphere; this mechanism involves only phosphates, magnesium, and water. Organic chemistry is involved at this point only to the extent that the phosphates attach to the ribose, which passively undergoes binding or hydrolysis of the phosphoryl groups, thus without playing a chemically active role. The fact that it is possible to devise a reaction mechanism accounting for the mode of action of ribozymes does indeed show that the RNA molecule possesses all the necessary atoms, as well as adequate steric flexibility, to catalyze a chemical reaction. Moreover, the RNAs have the ability to hybridize with a complementary sequence of the substrate, thus decreasing the total free energy of the whole so formed by means of stacking interactions. The possibility of forming hydrogen bonds with the substrate also constitutes the basis for specific recognition of the latter.

The experimental demonstration that the same macromolecule can possess both catalytic properties and the ability to self-replicate throws an entirely new light on the processes involved in the appearance of life on Earth. It constitutes a major event in the course of research which up to that point had seemed to be destined to pure speculation on the crucial points.

Here it is tempting to draw a parallel with the logic that led astrophysicists to the Big Bang theory to account for the origin of the Universe. The similarities between two of the most exciting scientific and intellectual adventures of the 1980s are striking. In both cases, observations of today's world, which is accessible to experimentation, supplied the foundation for hypotheses which, because they relate to events in the past, cannot be tested directly by experimentation but which rationally suggest themselves strongly to researchers.

In the case we are looking at now, the discovery of ribozymes that have survived to our time makes highly plausible the idea that evolution could have gone through an early stage in which only RNAs existed. The ribozymes still present in today's living world can be considered to be "chemical fossils" which give some idea of the nature of the first metabolic reactions in a long-gone prebiotic era. The study of these living fossils, represented by molecules that evolved very little through the ages, and by highly archaic reaction mechanisms, supplies more information: it allows us to go back further through time than by looking for fossils in the classical sense which are the oldest traces of cell life on Earth. These "molecular fossils" open the door to the precellular epoch and allow us to glimpse how the pathways by which certain basic mechanisms of today's living world and certain other biological macromolecules, more "modern" than the RNAs that gave way to them, may have developed. If RNAs have catalytic properties, the scenario comes into clearer focus. Indeed, three of these major riddles have been solved.

CONSEQUENCES OF THE DISCOVERY THAT RNAs HAVE CATALYTIC PROPERTIES

The first consequence is the initial step toward changing the genetic message. Logically, the self-splicing reaction that RNAs are able to bring about should be reversible. Thus, if an intron is capable of self-excising, it should also, conversely, be capable of inserting itself elsewhere into an appropriate nucleotide sequence. Thus it may be assumed that a segment containing two introns on both sides of an "exon," a coding sequence of a nucleic acid, can be transferred from one RNA molecule to another in the manner of a transposon, a movable element of the genome.

Such a property would have conferred on RNAs an evolutionary attribute of the utmost importance: the ability to remanipulate and create new combinations of genes. In the normal course of things, RNA molecules are able (albeit slowly) to undergo evolution due to replication errors, namely mutations. But in fact transposons are a kind of equivalent, still very simple, of what sexual reproduction would later be, ensuring transmission of genetic elements from one organism to another. Recombination and sexual reproduction are powerful mechanisms allowing useful exons to pass from one replicative system to another. Thus we have the possibility that relatively simple chemical processes may have established the most fundamental phenomena of life at a very early stage.

The second consequence relates to the synthesis of peptides. Recent experimental data suggest that, because of their chemical properties, RNAs made the emergence of proteins possible. An examination of the catalytic properties of ribozymes shows that there is nothing to contradict the notion that the primordial RNAs may have promoted the polymerization of another kind of monomer—the amino acids. From the thermodynamic standpoint, such polymerization appears to be equivalent to that of the nucleotides: the stability of a peptide bond and that of a phosphodiester bond are substantially the same, because the free energies of hydrolysis of each of them are very similar. As a result, if ribozymes are able to create new phosphodiester bonds, they should also be able to form peptide bonds. There is nothing against the hypothesis that nucleic acids may, by a reaction mechanism similar to the one leading to RNA polymerization, have been able to facilitate the polymerization of amino acids by supplying a suitable catalytic center and sacrificing one phosphodiester bond to permit the formation of a peptide bond. Thus, the problem of the availability of an external energy source for producing activated amino acids would have been solved in a way still primitive but sufficiently effective.

This attractive hypothesis was developed in 1987, particularly by Alan M. Weiner and Nancy Maizels, based on the observation of certain RNA viruses which they considered to be living fossils, reminiscent of the ancient RNA world. Often, near the 3' end, the RNAs of these viruses have fragments containing a 3'-terminal CCA triplet, like the one found in modern tRNAs; they also resemble tRNAs either by their sequence or, more frequently, by their ability to react with proteins that recognize the tRNAs. The two researchers assumed that these fragments allowed the ribozyme replicases to identify the 3' end of an RNA to be copied and served as an initiation site for replication in the 5' to 3' direction, as is still being done today. Any replication system, however primitive, does need to recognize the molecule it copies as well as the starting point of replication. This is why RNA molecules that did not contain this "quasi-tRNA" 3'-terminal fragment would not have undergone effective replication and would have eventually disappeared.

Let us follow the argument presented in the form of a historical narrative. It is probable that the distinction between genomic and functional RNAs did not originally exist. Later, when the number of ribozymes capable of different kinds of activities increased, the presence in all these RNAs of the same 3' marker fragment would limit the variety of enzyme reactions in which the 3' end could be involved. The appearance of an RNA-endonuclease able to remove this fragment in a fraction of the population of RNAs of the same species thus played an important role in evolution: it made diversification of enzyme activities possible. Ablation of this sequence transformed the genomic RNAs into catalytic molecules. The "quasi-tRNA" fragments, now useless, were then degraded or possibly recycled, but could also have been reused as transfer RNAs in protein biosynthesis. So, Nature made economical use of waste. This fragment is still cut today by the ribonucleic component of RNase P, but the mission of this enzyme has been reversed: instead of releasing an enzyme RNA by removing the 3'-terminal fragment, its function now consists in producing a functional tRNA from a precursor RNA.

Weiner and Maizels assumed that certain variants of RNA-replicases could have acquired a particular activity consisting in joining an amino acid to an RNA by using a reaction mechanism fairly similar to that operating in the fusing of a nucleotide. This replicase would have a binding site for the 3'-terminal CCA of a "quasi-tRNA," allowing the latter to be replaced in the active site of the ribozyme. A nucleotide containing any N base would first be bound to the replicase as occurred in the poly(C)-polymerase activity of the *Tetrahymena* IVS. However, instead of the phosphodiester bond thus created then being attacked by an OH group of an RNA, it would instead be cleaved by the carboxyl of an amino acid located in the active site thanks to a special affinity due, for instance, to the basic character of this amino acid. On the other hand, the activated amino acid would then be loaded onto the "quasi-tRNA" in the usual manner by the hydroxyl group of the 3'-terminal A residue. In this case, it would lead to formation of a carboxylic ester bond. The tRNA aminoacylation process is still, roughly speaking, the same today.

We have seen that amino acids activated in this way may undergo spontaneous polymerization to yield polypeptides that are short but functionally more useful than peptides with any randomly formed sequence. Moreover, a ribozyme (formed *de novo* or resulting from duplication of a replicase) having two tRNA binding sites may have speeded up synthesis of these peptides by properly juxtaposing two aminoacyl-tRNA molecules. Indeed, it is assumed that, even during modern protein biosynthesis, formation of a peptide bond is not enzymatic, but purely chemical; in other words it results simply from an ideally favorable juxtaposition of the acting molecules. Such a ribozyme would already have had the characteristics of a protoribosome, although the manner in which it functioned would still have been very primitive. Synthesis could have been effected by a series of repeated association-dissociation cycles, in which the loaded tRNAs would be associated and free tRNAs dissociated.

Thus the hypothesis is formed that the protoribosome was nothing more than a structure capable of interacting with molecules such as tRNA, aminoacyl-tRNA, and messenger RNA, or it might even itself have assumed the roles of all the components in this system, except of course the amino acids. This system would then have diversified with time, due on the one hand of the appearance of new aminoacyl-tRNA synthetases, each capable of reacting with a given amino acid, and on the other hand to new protoribosomes able to distinguish among different aminoacyl tRNAs. It is possible that this recognition occurred because of a special sequence of bases in the tRNA constituting the precursor of the anticodon; it could be attached to a complementary sequence of the protoribosome which would itself have played the role of messenger. This would explain how the first rudiments of the genetic code fell into place!

Under the effect of selective pressure, the advantage of synthesizing longer peptides would then emerge and lead to replacement of this internal mRNA by an external messenger, a special RNA associated with the system and coding for many amino acids.

The Weiner and Maizels hypothesis was rounded off by Calvin K. Ho. This scientist suggested that the primitive tRNAs and ribozymal aminoacyl-tRNA synthetases were produced from the same precursor, with the synthetase composed of an intron excised by self-splicing of a pre-tRNA molecule. This simultaneous synthesis of tRNA and the corresponding synthetase would have ensured a balance between the quantities of substrate and enzyme; in this way it would have contributed to regularizing the peptide formation process.

Such a general hypothesis of protein formation, in which all the steps are based on reaction mechanisms observed today in systems that have been highly preserved through the ages, appears to be entirely plausible. There remains, however, one major question: how could selection be applied to such a system to ensure its evolution toward greater diversification and lead to today's extremely precise and sophisticated biological processes? How are we to account for the undeniable progress thus accomplished? Here we must assume the existence of a feedback mechanism: the product of a reaction must be able to influence the speed of the process which gave rise to it.

The fact that the system of biosynthesis would be informed of the usefulness of the product it created would favor the appearance of new molecules that supply an advantage or, conversely, that hinder the development of useless products.

Here again, we have recently had available to us the elements of a scenario to explain the action of such a selection very early on, perhaps as far back as the stage of the RNAs themselves, before the appearance of proteins. In the RNA world, self-splicing was governed by the conformation of introns (presence of single-stranded zones forming loops, and double-stranded regions). Only a few fragments of RNA endowed with three-dimensional complementarity could lead to stable functional molecules by means of splicing. Thus, the viability of the structural domains of RNAs, based on their chemical and thermodynamic stability, defined the structure of the first exons. It is estimated that the length of the "selectable" fragments is a hundred nucleotides or so: the tRNAs, which are the smallest biologically active polyribonucleotides, include about seventy-five residues and the exons of today correspond to thirty to fifty amino acids. It is likely that later, when the ribozymes started to catalyze peptide synthesis, the positively charged amino acids became preferentially attached to the negatively charged polynucleotides. Hence, the earliest aminoacyl-tRNA synthetases had to be specific for basic amino acids. Given the acidic and hydrophilic nature of RNAs, the most useful of the earliest polypeptides were basic and hydrophobic, since they could protect the RNAs and stabilize the lipid vesicles which were beginning to lay down membrane structures.

In general, it is believed that the earliest peptides had no enzymatic activity and that their only role was to attach to RNAs in the manner of cofactors modulating and stabilizing their conformation and hence the activity of the ribozymes and ribosomal RNA. Thus, a combination with peptides could have made a major contribution to the increase in the number of stable functional RNAs and the diversification of their catalytic activities. In this way, protein synthesis could have become more effective and increasingly dependent on RNA structure so that it would end up being coded by the RNAs.

So this is the stage at which the second important function of RNAs would have appeared: transmission of the genetic message thanks to their ability to store information. In a parallel development, peptides would have begun to drop their role as mere ribozyme cofactors and gradually acquire catalytic properties themselves. In this movement towards the formation of protein enzymes, which should henceforth be distinguished by the name "proteozymes," the acquisition of new functions appears to have been a priority; but, as in the case of the polynucleotides, the structural integrity of the polypeptides was a primordial condition of the process. Without this stability, a protein cannot function effectively either as an enzyme or as a structural molecule.

Among the peptides, a selection similar to that of the polynucleotides must have operated in favor of molecules capable of assuming a stable three-dimensional structure. This stability requirement imposed a length limit on peptides of about thirty to fifty residues. Later, in response to evolutionary pressure toward the creation of larger and more efficient proteins, longer and more complex genes would have formed by the splicing of RNA, fusing of the coding parts, and elimination of the introns. Next, polypeptide reactivity would be gradually established by point mutations in response to pressure from the environment. We should add that, as in the case of the ribozymes, metals must have intervened frequently in early proteozymes and that later the structural flexibility was able to improve enzyme activity.

There is some reason to think tha t, following the combination of the first peptides and RNAs, the first proteins to appear would have been those that acted on nucleic acids, their natural partners. Of particular importance was the appearance of reverse transcriptases which began to transfer the genetic message to DNAs.

The mechanisms we have just described, which were put in place very early in the course of evolution, may have operated in the "progenote," the hypothetical common ancestor of prokaryotes (simple cells without a nucleus) and the more complex eukaryotes which have a cell nucleus. "Mosaic" genes composed of dispersed exons separated by introns still exist today in eukaryotes, but not in prokaryotes. This would be the result of a difference in survival strategies adopted by these cell lines at the time of their divergence, about a billion years ago! The biosynthesis of proteins from mosaic genes is a complex and energy-consuming process, and as such is incompatible with rapid cell growth. It looks very much as if, by eliminating the introns, the prokaryotes chose the option of rapid growth by simplifying the process and making it more efficient. With the slower-growing eukaryotes, energy expenditure was proportionally lower and hence more tolerable and, to defend itself against environmental challenges, this line chose instead to organize itself into multicellular assemblages. Thus, in contrast to the prokaryotes, it preserved the great evolutionary potential conferred by the mosaic gene system. So, paradoxically, the most primitive genes are not specific to the primitive organisms, but are also present in the most evolved organisms.

Finally, the demonstration that RNA had catalytic properties led to the formulation of hypotheses about the emergence of replicative processes.

The appearance of peptides probably activated the setting up of the first such processes. Among the earliest biopolymers, only the polynucleotides could, because of their structure, replicate by autocatalysis. We have seen that, due to the complementarity of the bases, a negative copy of a polynucleotide can form in a medium containing monomeric nucleotides which constitute both the raw material and the energy source for such synthesis. Replication of this negative copy then produces a new copy of the original polynucleotide. So, this system of positive-negative strands appears to be an autocatalytic entity capable of forming self-reproduction cycles. However, it has been found experimentally that, in the absence of proteins, formation of polynucleotide chains in this fashion is not very faithful. It improves in the presence of polypeptides, even short ones composed only of a few types of amino acids, with low specificity and only moderate catalytic capacity. Nonetheless, such a rudimentary system cannot reproduce indefinitely. Lacking its own program, it runs the risk of fairly rapid degeneration of the original sequences and proves incapable of maintaining a durable "lineage" of molecular structures, and of evolving and improving.

To be more efficient, replication must be based on a system characterized by a superior degree of organization. In the 1970s, Manfred Eigen actually advanced such a hypothesis. A nucleic acid, he explained, may direct the synthesis of peptide chains with some efficiency. Reciprocally, these peptides can catalyze the synthesis of another polynucleotide chain during a subsequent replication cycle, and so forth. Several conjugated reproduction cycles of this kind could constitute a new entity of a higher order, or "hypercycle," able to self-multiply according to a program contained in the nucleic acid. This would be the simplest system capable of reproducing itself according to a program and the first to be able to undergo evolution and dynamic selection, as the most advantageous mutants would multiply the most efficiently. Thus, according to Eigen, the hypercycles would have opened the way to the emergence of the first simple living organisms.

How could hypercycles have appeared? Can we fall back on chance when we are looking at ordered systems consisting of several components that have achieved a certain degree of complexity? Highly unlikely! On the other hand, such systems, which are ordered in space and time, could come about by spontaneous self-organization through the "dissipative structures" that we will discuss in the next chapter.

III

FROM THE SIMPLE TO THE COMPLEX

HIGHER-ORDER SYSTEMS

The discussion on the development of simple systems has already taken us quite a long way toward the formation of new prebiotic molecular species that are rather elaborate and are endowed with various kinds of chemical activity. How could the organization of matter have been able to evolve spontaneously to higher levels? To state the question differently, how can chaos generate orderly systems?

Work by the physical chemist and Belgian Nobel Prize winner Ilya Prigogine has opened up new avenues of research and perhaps supplied a few solutions. Let us remember that, according to the second law of thermodynamics, an isolated chemical system reacts according to the affinity of the molecules of which it is composed and to environmental conditions; during this process, the free energy of the system decreases and the entropy increases. The entire system tends toward a state of chemical equilibrium, an "attractive" state in which the free energy reaches a minimal endpoint and entropy is at its maximum.

In dealing with open systems communicating with the outside environment, we cannot confine ourselves to looking at changes in entropy produced by the system itself: we must take into account the energy transported toward the environment. Also, entropy production allows us to distinguish three domains in thermodynamics: that of equilibrium, studied largely in the 19th century, and the two domains of nonequilibrium described more recently, close to equilibrium (linear thermodynamic) and far from equilibrium (nonlinear thermodynamic).

In the state of equilibrium, production of entropy is zero; the same applies to the rate of chemical reactions and other irreversible processes or "flows" which nourish them (heat transport, diffusion of matter, etc.), as well as the forces which cause these flows (chemical affinities, temperature or potential gradients). In the state of near equilibrium, where thermodynamic forces are weak, flow is a linear function of force; thus for example heat flow is proportional to the temperature gradient and diffusion flow is proportional to the concentration gradient of the diffusing substance. When a system is under conditions in which the forces are other than zero and continue to produce low rates of flow, it cannot achieve the equilibrium associated with zero entropy production; it then stabilizes its state at the minimum level of entropy production and dissipation. These are the key concepts of the linear thermodynamics of near-equilibrium states.

So, in the same manner as equilibrium thermodynamics, stable, predictable behaviors of systems are described which tend toward a minimum (low or zero) level of activity compatible with the flows to which they are subjected.

On the other hand, when they are remote from equilibrium, the flows are not linear functions of the forces. Here we enter the sphere of "nonlinear" thermodynamics where it is not possible to define an attraction state, the stable endpoint of irreversible evolution. The instability of the steady state then triggers a phenomenon of spontaneous self-organization. Prigogine illustrates this situation by a physical example of instability called "Bénard instability." An instability is created in a horizontal layer of liquid by subjecting it to a vertical temperature gradient: the lower boundary is brought to a temperature slightly above that of the upper boundary. The molecules of liquid then undergo disorderly agitation and transmit small quantities of heat upward. But if the temperature differential is increased, a threshold appears above which the steady state, at which heat is transported by diffusion, becomes unstable. The molecules begin to move coherently and this macroscopically observable convection phenomenon accelerates the transport of heat; this means that it increases the production of entropy in the system. A new supramolecular order is thus stabilized by energy exchange with the environment.

Structures of this kind that spontaneously access a degree of higher order are called "dissipative structures" by Prigogine because they produce excess entropy and dissipate it into the environment. This term combines the idea of order with that of waste and expresses a fundamental fact that had hitherto gone unnoticed: remote from equilibrium, dissipation of energy becomes a source of order.

Systems that are out of equilibrium are also encountered in chemistry. Disruptions in thermodynamic stability are observed in autocatalytic chemical processes during which one of the compounds present in the reaction medium increases the rate of the reaction which gives rise to it, or, conversely, decreases the rate of the stage during which it is consumed. When the concentration of certain reagents exceeds a particular threshold, the reaction organizes itself spontaneously into a periodic process and proceeds alternately in one direction or the other.

The earliest observations of such processes were made in the 1950s by B. P. Belousov and later developed by his compatriot, the chemist A. Zhabotinsky. Reactions of this type are called "Belousov-Zhabotinsky reactions." Oxidation of an organic acid by potassium bromate in sulfuric acid in the presence of a catalyst (cerium or manganese ions) is a good example of this: an oscillating regime becomes established in a homogeneous medium and may be maintained by continuous introduction of fresh reagents. When the reaction medium is given the shape of a thin liquid layer in which the reagents cannot mix uniformly, an additional parameter is introduced: diffusion of reacting substances. The reaction then proceeds along ordered lines not only as a function of the time parameter but also as a function of space, with waves of concentration and chemical reaction forming in the liquid that are even more spectacular when the medium is colored.

Since the first results of Belousov and Zhabotinsky, other oscillating reactions have been discovered; to date, thirty or so chemical oscillators have been discovered which alternately produce and consume substances that participate in the reaction. The theoretical aspects of such dissipative dynamic systems have also been worked out. It should, however, be noted that "nonlinear" reactions, whose effect (formation of the reaction product) feeds back to the cause (synthesis process), are somewhat rare in chemistry. In the world of living organisms, on the other hand, "feedback loops" based on autocatalysis or self-inhibition (presence of a substance respectively accelerates or blocks the process of its formation), which are classical mechanisms of metabolic regulation, condition chemical instability remote from equilibrium and lead to nonlinear systems. Living things function mainly in the domain remote from equilibrium, where entropy-producing processes, which dissipate energy, contribute to creating order.

Although this was not their original goal, the work of Prigogine and his adherents thus made a major contribution to the analysis of the chemical functioning of the living world, kept remote from equilibrium by constant flows of the energy and matter that feed it. Because of this work, it is possible to understand the phenomena of self-organization of matter within the framework of the general laws of thermodynamics.

COMPLEX STRUCTURES

The extraordinary diversity of proteins in today's world of living organisms also results from a process already observed in the prior stages of the organization of matter: first, there is a relatively limited number of certain structural elements; then structures that are more complex and highly diversified form through different combinations of these basic elements. This is the organizing principle of today's proteins, whose origins date back to the archaic world of the RNAs.

This process of creation can be observed in two ways. If we study the composition of a family of proteins, we often see the presence of similar elements and can try to deduce the history of the slow formation of these macromolecules in the course of evolution. But we can also observe the appearance of new proteins in a process which, by contrast, is very rapid, namely the formation of antibodies (immunoglobulins) by the immune system.

Certain proteins form simply by joining together a large number of similar units. Thus, a fibrous protein, collagen, is composed of three peptide chains; these chains form helixes, each of which has about 1000 amino acid residues distributed in repeated sequences of adjacent tripeptides. The first amino acid in each of these sequences is always a glycine. With this monotonous structure, collagen resembles artificial polymers. A protein of this kind is formed from a small primordial gene by a simple genetic mechanism consisting of multiple duplication of genes. Functional proteins generally have a more varied structure.

During the last few decades, the development of the new methods for studying proteins mentioned above has allowed a large number of these macromolecules to be analyzed. It has been found that polypeptide chains containing about 200 amino acid residues are in fact organized into at least two regions of autonomous structures; they have the appearance of a complete protein molecule, but are linked by a peptide chain to one or more other regions. These structural domains usually have receptor properties for other molecules and, if the protein is an enzyme, the active site is almost always at the interface between the domains. It looks as if, in the course of evolution, these proteins were formed by fusion of two or more "protoproteins" by a mechanism involving gene fusion.

These observations appear with particular clarity in the work of Michael G. Rossmann on the structure of dehydrogenases which belong to the family of glycolytic enzymes. These proteins are involved in the fundamental processes of life as it existed even before the separation of the prokaryotes and eukaryotes. Their amino acid sequences, like their tertiary structures and catalytic mechanisms, have been preserved through the ages. Since their early development during a very remote epoch, they have undergone no noteworthy changes.

Rossmann found that in a number of dehydrogenases functioning with the assistance of the coenzyme NAD (nicotinamide adenine dinucleotide) the regions attaching the latter had very similar structures and coenzyme-binding properties, while their catalytic region differed from one enzyme to another.

He concluded that the binding sites of the nucleotide had descended by duplication from a common ancestor and that later they had bonded to different catalytic regions to form distinctive dehydrogenases. By combining different functions in one protein molecule, such a process would have been able to produce new proteins rather easily. This would be compatible with the observed rate of evolution on Earth, while point mutations could not account for this all by themselves.

The hypothesis of exon shuffling thus explains the existence of several individualized domains in most proteins, as well as the repeated use of similar structural motifs in different proteins. For example, a group of eukaryote proteins including, among others, fibrinolytic proteases acting on blood clots and those involved in blood coagulation. All these proteins are composed of a certain number of similar functional domains assembled in different ways. Their function is to ensure the binding of the catalytic domains of these proteins to other molecules or to macroscopic structures. Thus, certain areas facilitate the binding of coagulation enzymes to their cofactors or to membranes; for example, the domain that binds calcium becomes attached to the membranes by forming a "calcium bridge" with the phospholipids. The amino acid sequence of this domain is highly preserved in all the proteins that contain it; in other words, it has changed very little in the course of evolution. The same applies to the other domains which have major structural similarities and are probably differentiated versions of one and the same basic structure; slight variations ensure their specificity. Fibrinolytic enzymes can use fibrin as a substrate to bring their catalytic domains into position. Fibronectin has several binding functions and can use several domains to attach itself to a variety of macromolecules (fibrin, actin, collagen, heparin, etc.). This last example shows that the combined intervention of several elements can ensure binding with great affinity.

Studies of this kind dealing with the structure of protein families enable hypotheses to be advanced regarding the way in which these macromolecules were gradually created during the course of evolution. Despite the complex structures of the enzymes we have just been discussing, it has been possible to draw a genealogical tree indicating the probable evolutionary chronology of the formation of these proteins by successive incorporations of the various domains.

The functioning of the immune system supplies another illustration, and a very spectacular one emerging from recent discoveries, of the possibility of creating a practically infinite number of new proteins by a combinatorial and economical strategy from a comparatively small number of structural elements.

In response to a stimulus that appears in the form of an antigen, the immune system very quickly triggers mechanisms similar to those which, under the effect of environmental pressure, modified and reorganized the genome over an infinitely longer time scale.

Described very schematically, the basic structure of antibodies is a unit containing four peptides: two "heavy" (H) chains and two "light" (L) chains connected by S-S (disulfide) bridges. Most vertebrates have two types of light chains designated by the Greek letters κ and λ and several types of heavy chains (μ, γ, α, etc.) which determine the class of immunoglobulins (IgM, IgG, IgA, etc.) and their functional characteristics. Each chain encompasses one region with a variable sequence (V_L or V_H) and another region which remains constant in a light or heavy (C_L or C_H) type of chain. The whole forms a Y, each of whose two arms has an antigen-binding site; this site is formed by the regions V_L and V_H and the diversity of the $V_L - V_H$ combinations is the basis for the diversity of antibodies.

The genes of mammalian antibodies exist in the form of small dispersed segments. Thus, for example, the variable region of the light chain κ can be encoded by several hundreds of V segments and four junction segments J, while the constant region is encoded by a single gene C. The genes of the heavy chains are similarly organized, but their variable region is encoded, in addition to V and J type segments, by several diversity segments D (which can be as many as twenty in number). The process leading to immunoglobulins is a two-stage one. First of all, during development of the cells that produce them, namely the B lymphocytes, a DNA reorganization leads to the functional gene that controls the expression of the antibody. In the course of this recombination, one V segment and one J segment of a light chain or a V_H, J_H, and D_H chosen at random fuse by means of enzymes which at the same time destroy the intermediate DNA fragments. In a second stage it is this reorganized DNA which is normally transcribed into pre-mRNA and, after maturation, the messenger is translated into an antibody—a protein bound to the outside of the cell membrane as a receptor recognizing a given antigen. In this way, a very large number of families or "clones" of cells form, each of which is capable *a priori* of responding to a particular antigen. The presence of this antigen triggers synthesis of the antibody which is no longer bound but is secreted and circulates in the blood. The first stage, during which the DNA undergoes a specific modification, is unusual and remains characteristic of the immune system alone even though it goes back for several hundred million years and is thus very ancient from an evolutionary standpoint.

The functioning of the immune system provides a striking illustration of the extraordinary diversity of assemblies that can be put together on a low budget from a limited number of individual parts. It is true that in this arena the needs of the organism are immense: it must respond to an infinite variety of antigens with unpredictable structures, including brand-new products of chemical synthesis. This system was set up because no genome can respond quantitatively or qualitatively to such a demand. The organism has no ready-made genes, but it does have the wherewithal to manufacture a multitude of genes by somatic recombination of DNA from dispersed segments, and the diversity of the genes formed corresponds to the product of the number of component segments. Thus, the combinatorial association of 250 Vκ segments with four Jκ segments can yield 1000 variable regions of a κ chain. For heavy chains, the genes formed by the combination of V_H, D_H, and J_H segments would reach 5000 to 10,000 in number. If all the V_L-V_H combinations were possible, between five and ten million different antibodies could be constructed from less than 400 genetic segments. Two other antibody diversity factors are involved. The first results from the relative inaccuracy of the DNA splicing mechanism, particularly where heavy chains are concerned: the ligation has a margin of error of several base pairs and, in certain cases, extra base pairs become inserted when segments are fused together. A second factor is the frequency of somatic point mutations in developing

B cells. It has been established that the rate of mutation amounts to a change in the V region every 3 to 30 cell divisions; thus this rate is several orders of magnitude greater than that of eukaryote gene mutations. Because of these supplementary factors, antibody diversity increases at least a hundredfold and, if combinatorial mechanisms alone can generate ten million antibodies, the total number would thus exceed one billion.

COMPLEX FUNCTIONS

When they gave birth to proteins, RNAs acted like sorcerer's apprentices. We saw that the first peptides served the ribozymes by increasing their stability, acting as cofactors to improve the performance of reactions induced by RNAs. But polypeptides, going beyond this modest initial role of RNA auxiliaries, developed a more complex chemistry endowed with major evolutionary potential. Hence the extraordinarily sophisticated chemistry of life that we observe today.

So let us now continue to follow the developments in chemical processes which were essentially due to the appearance of proteins.

To understand this evolution, we should analyze the current situation which is its outcome. Let us therefore systematically take another look at how the catalytic activity of today's enzymes is exercised and what factors it enlists.

• An enzyme binds to the substrate at the active site in such a way that the bond of the substrate about to be transformed is located near the catalytic group and is properly oriented with respect to it; this makes it easier to achieve the transition state.

In the transition state theory, the rate of a reaction depends, as we have seen, on the difference in free energy of activation ΔG^{+} between the starting products and the activated complex. This value is determined by the difference between the "enthalpic" term ΔH^{+} corresponding to the variation in binding energies and the entropic term $T\Delta S^{+}$ which expresses the variations in the degree of freedom of the system:

$$\Delta G^{+} = \Delta H^{+} - T\Delta S^{+}$$

Hence these two types of variations have opposite influences on the rates of reactions. For example, if by adding an acid capable of liberating a proton H^{+}, a negatively charged transitional state is stabilized, the reaction is facilitated by decreasing the ΔH^{+}; on the other hand, however, ΔS^{+} also decreases because the binding of the acid reduces the degree of freedom of the system. If several independent catalyst molecules are to be involved in the reaction, binding of each one of them may contribute to reducing the ΔS^{+} counteracting the catalytic power. But if, as in the case of an enzyme, all the catalytic groups are located in the same molecule, the drop in entropy is smaller because the freedom of movement of a single molecule is lost.

• An enzyme can combine with the substrate to form an unstable covalent intermediate which undergoes the reaction more readily. The reaction takes place in two stages with a low energy barrier, the sum of which is less than the activation energy of the overall noncatalyzed reaction.

• Various enzyme functional groups are proton donors or acceptors and can play the role of general acid-basic catalysts.

• An enzyme can create a steric constraint in the substrate which weakens the bond to be broken.

• The enzyme can form a hydrophobic cavity around the active site, thus favoring polarization (separation of charges) of its catalytic groups and polarizability of the substrate bond that is to undergo the reaction.

It would be of interest to discover the chronological order in which these various factors may have entered the picture during evolution. Here too, a few hypotheses may be suggested.

It is certain that catalytic activity appeared very early under the conditions of the primitive Earth: even very small organic or inorganic chemical entities are able to play this role. Thus, H^+ and OH^- ions arising from dissociation of a water molecule constitute specific catalysts, acid and basic, respectively; these catalysts, the most universal, are capable of speeding up reaction rates in proportion to their concentration.

Later on, the formation of slightly more complex acid or basic molecules may have given birth to general catalysts, proton donors or acceptors in terms of the theory developed by the chemists J. N. Brönsted and T. M. Lowry. At a neutral pH, where the concentration of H^+ and OH^- ions is very low, general acid-base catalysts can speed up chemical reactions which, in their absence, could take place only in strongly acid or basic media. It is mainly the peptides that supplied a wide variety of new catalysts of this type.

The efficiency of such catalysts depends on their acidic or basic strength, in other words on their proton-dissociation constant. But it appears that the speed with which the acid or base can give off or accept a proton has an important part to play. There is one amino acid, histidine, which not only can act as a donor and acceptor of protons at physiological pH levels, but in addition its protonation and deprotonation rates, substantially equal, are very high (half-transfer time less than 10^{-10} seconds). Already in the proteinoids, the presence of histidine contributes a great deal to the catalytic activity of these primitive polypeptides, and we find that this amino acid, although it is rather rare, is involved in the catalytic processes of numerous present-day enzymes.

But in the early stages of evolution, catalysts, which were tiny at the time, could have had no substrate specificity. They must have been capable only of accelerating a number of reactions to a somewhat moderate degree and with no great discrimination.

To become more efficient, the catalytic groups had to be inserted into macromolecular structures. These structures allow for the formation of active sites, assemblies of several catalytic groups acting in concert and improving chemical processes. The same macromolecular structures are also essential for ensuring substrate specificity.

Ribozymes, the earliest biological catalysts, often reached considerable size. However, their ability to react was limited by two factors. The first had to do with the narrow range of chemical strategies available to them: water, metals, phosphates, and electrostatic attraction. The second had to do with the difficulties nucleic acids have in adopting complex tertiary structures that facilitate catalytic processes; as we have seen, the steric factor is mainly responsible for accelerating enzyme-induced reactions. Polynucleotides were able to undergo intramolecular reactions between residues that are distant along the chain but spatially close by folding of the polymer, facilitated by base pairing in some of its segments.

On the other hand, the ability to hybridize with a complementary sequence of another RNA, a modest first step in substrate specificity, had the effect of promoting intermolecular reactions. The flexibility of the polynucleotide chains and the possibility of pairing between chains can therefore be regarded as chemical assets of RNAs resulting from their primary (nucleotide sequence) as well as secondary structure (double-stranded regions and loops). In this way, ribozymes were able to develop a chemistry that was already relatively varied but probably reached its limits fairly quickly because of the difficulty RNAs experience in forming higher-order structures.

In polypeptides, these two obstacles are removed. In fact, peptides introduced new functions into evolution: those of organic chemistry. This made it possible for chemical processes to diversify a great deal while becoming more energy-saving. Moreover, unlike nucleotide polymers, amino acid polymers can adopt tertiary structures of an infinite variety; the residues can interact with each other, with the molecules of the solvent, or with other dissolved substances.

In proteins, amino acid residues are hence able to be brought together to form structures capable of attaching to the substrate through numerous non-covalent bonds which ensure great specificity. Also, the catalytic groups can be disposed sterically so as to form complex but highly efficient reaction sites, endowed with great evolutionary plasticity and open to subsequent improvements. We can see that the development of the chemical performance of living matter adopted the same strategy as that of molecular structures: the combination of a limited number of simple chemical functions, or "functional building blocks," judiciously assembled, led to catalytic systems of extraordinary efficiency. Today we thoroughly understand the action mechanisms of several enzymes; we cannot but admire the degree of precision which the evolution of certain kinds of chemical machinery was able to achieve.

Thus, the first "attempt" by Nature to base the chemistry of life on nucleic acids did not go very far. Long polynucleotide sequences are very suitable for storing genetic information, but their simple linear chemistry does not fit in very well with the three-dimensional structure of living cells and their components. Amino acids, on the other hand, more numerous and chemically more diversified than nucleotides, produce polymers capable of assuming an infinite variety of spatial configurations. Each sequence of amino-acid residues corresponds to a given tertiary and possibly quaternary structure that has minimal energy and is therefore thermodynamically stable. So if the sequence of residues in nucleic acids apparently determines only the sequence of amino acids in the polypeptides derived from them, this linear information in fact indirectly contains at the same time three-dimensional information on the spatial structure of the products of translation. Hence, the appearance of proteins introduced a new dimension, in both the literal and the figurative sense, in the development of the chemistry of life. These new macromolecules, capable of assembling in space effective chemical tools to form specific and efficient catalytic sites, opened the gates to virtually boundless evolutionary prospects that enabled the organization of matter to make decisive and revolutionary strides. There were many steps still ahead and numerous organizational crises to be surmounted, but the advent of proteins created a chemical basis that established an adequate foundation for these future developments.

IV

THE "ANIMATION" OF MATTER

We can now imagine the next corner turned by evolution once the informational and catalytic molecules as well as membranes were in place. Of course, this is not unbridled imagination but a coordinated set of hypotheses representing the extrapolation of observations and confirmed experimental results.

The first multimolecular units must have functioned like anaerobic heterotrophic protocells, using the surrounding organic compounds as the source of building materials and energy. Once it was mobilized to feed the biosphere, the organosphere or "primeval soup" gradually became depleted. It may well be that the environmental pressures brought about by this food crisis triggered the appearance of autotrophic cells capable of feeding on carbon dioxide and atmospheric nitrogen (however, according to another hypothesis, autotrophic cells may have appeared first). Photosynthetic cells using light as an energy source may have been the response to the dearth of energy. Thus the atmosphere gradually became enriched with oxygen, eliciting the appearance of aerobic cells that not only could withstand this pollution but could even turn it to their advantage by developing respiratory and oxidative phosphorylation processes capable of extracting energy from nutrient molecules more completely. The endosymbiosis of small photosynthetic or heterotrophic prokaryote cells—the precursors of chloroplasts and mitochondria—with large anaerobic prokaryote cells led to eukaryotes: a new and very important step forward on the path of evolution. All these developments already belong to the field that we call purely and simply "biology."

But where are we to draw the line between the inanimate and the animate? Can we pinpoint the exact instant at which "life" appeared in this continuous process of the organization of matter? The answer to this burning question assumes that we are capable of defining the principal feature of life that allows a clear distinction to be made between the living and the nonliving.

It is generally accepted that the organizational status that characterizes living organisms is composed of their self-organizing faculty (metabolism) and their ability to self-replicate. So it is to these two characteristics that we must now turn.

METABOLISM

How did metabolism, the flow of matter and energy that passes through organized beings, appear in the course of evolution? What lessons may be learned by comparing ancient forms with the organization of today's living beings?

Let us turn once more to the earliest stages of chemical evolution which, beginning with small primordial molecules, led to compounds of greater complexity which, in turn, were able to unite to form larger assemblies: coacervates or microspheres. Within a space delimited by a quasi-membrane, such structures could have enclosed macromolecules that exhibited a certain measure of catalytic activity. It is here that we see the first signs of primitive metabolism. Small molecules (sugars and amino acids) can actually penetrate inside a coacervate and be converted there into another substance which then diffuses outside. Such a unireactional metabolism can be observed experimentally and can also be modulated. In this way, coacervates containing a given catalyst can be formed at will. Droplets containing an enzyme such as, for example, a phosphorylase, when placed in a phosphorylated glucose solution, accumulate inside a polysaccharide of the starch type and expel the inorganic phosphate into the surrounding milieu. If, in addition, they contain a hydrolytic enzyme such as an amylase, the starch formed is then broken down into small sugar molecules (maltose) which are discharged outside; the coacervate is then endowed with a bireactional metabolism. Numerous other functional coacervates have been designed. These are, of course, only models made artificially with the aid of present-day enzymes which are certainly more efficient than the catalysts that may have existed in a long-gone era. We should note, however, that the appearance of catalytic activity in microspheres or coacervates results in these objects being transformed from static, closed systems into open steady-state systems whose matter and energy flows are in balance with the environment. Thus their stability no longer depends on thermodynamic equilibrium in a maximum entropic state, as in the case of closed systems, but on the constancy of the material flow over time. Free energy enters and leaves at the same rate in the case of open systems, while it is constant in closed systems. The entropy of open systems, on the other hand, although it too is constant, does not reach its maximum: an open system is constantly discharging entropy into the surrounding medium. It remains at a relatively low level by absorbing low-entropy molecules, transforming them, then dumping the waste back outside because the waste is in a more disorganized state of matter and hence at a higher entropy. There can be no doubt that this operating modality already corresponds to that of a "living" cell, an open system. The fact that these manifestly inanimate objects are capable of some degree of metabolic activity does suggest that metabolism cannot be deemed an exclusive attribute of life.

Turning the question around, let us now examine the organization of present-day living organisms and see whether there are any living systems that have *no* metabolism. This comes down to asking what is the minimal organization of matter that merits characterization as "living." This reasoning may be compared to that of Dalton, who, after thinking about the divisibility of matter, arrived at the idea of the atom; in the same way, Hauy, following a similar line of thinking in the field of crystallography, came up with the idea of the elementary crystal lattice conceived of as the smallest element of a crystal.

It is sometimes maintained that the cell is a "bioatom," arguing that below the order of magnitude of the micron there is no way a structure with a sufficient degree of complexity and precision to be alive can form. Others believe that elements such as bacterial spores, 0.12 thousandth of a millimeter in size, are at the lower limit of the so-called bioatom dimension.

But it is possible to move continuously and gradually still lower down the scale of magnitude by considering the viruses.

It is not always easy to draw a clear line between the largest viruses and the smallest primitive parasitic bacteria. Let us consider, for example, the infectious agent of psittacosis, or parrot disease, which was formerly believed to be a virus; it is now classified in the family of the *Chlamydia*, which are very primitive prokaryotes. This organism is tiny (0.2 to 0.5 thousandth of a millimeter) and contains DNA and RNA as well as a few enzymes needed for biosynthesis of some of its components; but as it has no ATP-ADP energy system and is lacking several fundamental enzymes, it cannot be cultivated in the absence of its host cells. At even simpler degrees of organization, the parasitic nature of the organism becomes accentuated. While the large viruses can contain a phospholipid membrane, a few hundred genes, and a few enzymes, the most rudimentary of them possess only a few genes and a single type of protein. At the very borderline, the "retroviruses" would be only the "free" form of a movable genetic element issuing from the hereditary patrimony of the cell, which would have acquired extracellular autonomy and infectiousness. Yet, despite the poverty of its constitution, the virus cannot be considered merely a chemical substance: viruses are really and truly living objects. The structure and reciprocal arrangement of their components are by no means random, but inscribed in their genomes. The furthest we can go is to regard this case as representing the extreme degree of parasitism: not only do these simple viruses lack a metabolism of their own, but they depend, even for their reproduction, on their host organism which supplies the necessary biosynthetic machinery and supplies of materials and energy.

So when we review the simplest life forms, we see that there is actually no clear line of demarcation between cells and viruses. When we move down the scale of organisms, we see on the contrary a smooth continuum, from autonomous unicellular creatures possessing all the biochemical equipment needed for survival to the more primitive organisms that have an increasingly impoverished inherent organization and an increasingly rudimentary metabolism, and which make up for these defects by increasing parasitism. In all cases, however, we say that we are dealing with "life."

So, on the one hand we can say that metabolism can be manifested in inanimate systems and on the other that it may not be present in living objects. It cannot therefore be considered to be the essential characterizing feature of life.

REPLICATION

We have seen the evolution of matter lead to the formation of multimolecular aggregates interacting with the surrounding environment. We may assume that environmental pressure has continued to be exerted on these open systems. Thus, if changes in catalytic activity occurred, only those which conferred a greater ability to survive on these structures must have been preserved at the expense of the other, less "successful" systems. Such a gradual evolution in molecular complexes that kept their structures in memory implies the existence of mechanisms that allow them to replicate identically, with durable changes in structure only by way of exception. In other words, biological macromolecules must have acquired the ability to replicate and self-multiply. We may thus assume that this phenomenon, which prefigures the ability of cells to reproduce, must be considered the essential attribute of life and marks the transition from inanimate matter to living organisms, and from "chemical evolution" to "biological evolution."

Yet it must be noted that the ability to replicate and self-multiply is not confined to nucleic acid-protein systems. It is a far more generalized property which can also be observed in certain mineral systems.

Armin Weiss aptly summarized the main characteristics of replication.

• Any system can be considered "replicative" if it is able to undergo self-multiplication during which a piece of information can be transferred from the template to the copy.

• This information is always carried by macromolecular systems. Prolonged storage of several items of information in small, isolated molecules is not possible.

• The information thus propagated generally concerns catalytic structures; it may be direct as in the case, described below, of certain mineral systems where the template carries catalytic sites directly and transmits them to the copy; or indirect, encoded as in DNA which transmits instructions for synthesis of catalysts.

- Replication must be precise. Occasional errors may quantitatively or even qualitatively change the catalytic capacity transmitted. Moreover, replication errors may slow down or, conversely, accelerate the course of subsequent replications. In the former case, erroneous replicas are eliminated; in the latter, the modified information becomes predominant.
- In a replicative system, it must be possible for template and copy to separate, with the copy in its turn becoming the template.

Now, such a reversible separation mechanism exists in the case of certain minerals, particularly the "montmorillonites," silicoaluminates with a lamellar structure. In the presence of water, intercrystalline swelling occurs due to the penetration of water molecules between the parallel layers of the montmorillonites; these layers peel apart, but their distance is inversely proportional to the concentration of electrolytes in the water. Above a certain concentration limit, the crystal disintegrates into independent flakes. But in a culture medium that contains the appropriate ingredients, each of the individual flakes can constitute a center of nucleation guiding the formation of exact replicas. Such a self-multiplication system can be brought about by freezing-thawing cycles or by alternating periods of rain and drought. However, at a given electrolyte concentration, lower than the threshold value triggering complete disintegration, the interstices between the layers enlarge only up to a given size, allowing nucleation and growth of new flakes in the interlayer region. It has been established that montmorillonite layers can undergo elastic deformation; the newly forming layer gradually pushes back the limit of elastic deformation of the parent layers. Such intercalating synthesis is thermodynamically favored and transmission of information on structural irregularities in the matrix is more efficient.

This replication mechanism can probably be considered a primitive two-dimensional analog of the one-dimensional replication of DNA.

What is the nature of the information transmitted by the replicative system of montmorillonites? It consists of local modifications in structure which are transferred from the templates to the copies. Substitution of the tetravalent silicon ion Si^{4+} by trivalent aluminum Al^{3+}, or Al^{3+} by magnesium Mg^{2+}, creates not only an excess negative charge but also triggers the formation of "Lewis base" sites. These bases are chemical entities endowed with catalytic activities which participate in numerous reactions. Thus the surface of a montmorillonite can be considered a primitive model of a multi-enzyme complex in which the various catalytic activities are partly interdependent. These activities are often selective as well: various compounds, such as amino acids or nucleic bases, for example, are selectively absorbed at the silicate surface. Thus, adenine is effectively adsorbed on differently charged montmorillonites, whereas thymine is not adsorbed, whatever the charge density. Is information faithfully transferred from the templates to the copies during intercalating self-reproduction in the case of montmorillonites? Actually, the margin of error appears to be relatively broad. This "deficiency" is attributed to the fact that the information vector, two-dimensional in this case, is less advantageous than a one-dimensional system such as DNA. Indeed, one of the major problems in replication resides in the identification of the starting point of replicative synthesis. While in the case of a one-dimensional template it is simply at one of the ends, in the case of a two-dimensional vector it can be at any point on its surface.

The frequency of errors accelerates the evolution of the system; changes in catalytic capacity and selectivity may increase or decrease the rate of self-multiplication. Thus, for example, if a "mutant" acquires a new catalytic activity, it may lead to the synthesis of a larger-size product; now this product may attach irreversibly to the surface of the catalyst, thus preventing its replication. In other cases, the "mutant" may have an abnormally high charge;

potassium ions K^+ then attach selectively to this site and in this way alter the dilation properties of the interlayer which can no longer achieve the necessary thickness to continue synthesis by intercalation.

It is therefore clear that certain clays possess the essential attributes of replicative systems containing molecular information. One may thus hypothesize that these inorganic compounds, abundantly present on the primitive Earth, may have played the role of mineral "quasi-genes" by assisting in the formation and replication of organic molecules, up to the point when these molecules, having acquired an independent ability to self-multiply, no longer needed to use mineral vectors for replication. This hypothesis of the mineral origin of life is owed in particular to Alexander G. Cairns-Smith. According to this scientist, the first steps in the prebiotic organization of matter were accomplished with the assistance of clays.

So we come down to the fact that, like metabolism, the ability to replicate also cannot be a sufficient criterion for distinguishing living from nonliving. Nor is it a necessary condition for considering a system as belonging to the living world. In a series of living objects arranged in order of increasing complexity: organelle, cell, organ, complete multicellular organism, this property appears and disappears periodically: the organelle and the isolated organ cannot multiply *in vitro*. Likewise, viruses do not have the ability to self-multiply in the strict sense of the term, but depend for reproduction on the host organism.

CONCLUSION

When we attempt to define life, or living, we immediately come up against a fundamental and apparently irreducible paradox: living organisms are composed of inanimate molecules.

An enzyme, for example, is a chemical able to catalyze a given reaction; it very efficiently fulfills its mission inside the cell, which could not live without it. But this same enzyme, taken out of the cell and purified, reveals its nature as a purely chemical object which can be manipulated like any other substance provided the precautions imposed by its complex structure are taken. Put it in the right artificial environment and, *in vitro*, it will keep right on doing the only thing it knows how to do, which is to catalyze "its" reaction like a robot, although there is no need for it to do so.

Must we then say that "life" is the interaction of all the inanimate components of this whole? In other words, that nothing is alive in a cell except the whole of it? But these interactions, complex and logically organized in time and space though they may be, obey exactly the same physical and chemical laws as inanimate systems. The apparent increase in the order of matter observed in biological processes does not in any way contradict the second law of thermodynamics because it is always accompanied by an increase in entropy elsewhere—often that of surrounding water molecules.

The sequencing of these processes is inevitable, just as inevitable as the attraction of ions with opposite signs in a solution. Likewise, when we consider most of the affinity constants (enzyme-substrate, hormone-receptor, certain proteins-nucleic acids) we can see why these molecules cannot help but combine rapidly and efficiently. When it comes in contact with the active site of the corresponding enzyme, the substrate finds residues of essential amino acids ideally disposed to react; the reaction will be clean and quick because under these conditions it cannot be otherwise.

However, due to the extreme complexity and perfect organization of biological phenomena, and although biomolecules obey the general laws of physics and chemistry, it is sometimes assumed that their interactions are governed in addition by principles of a higher order—by a certain special molecular logic of living things. It is true that the complexity of the mechanisms governing molecular interactions increases considerably with the level of organization of matter. But do we have to assume that there is such a special logic, in the light of what we now know of the origins of life?

The evolution of matter, based on perfectly objective and universal factors such as the chemical properties of atoms and molecules as well as the laws of thermodynamics, was inescapable. What is more, we have no reason to believe that it could have occurred only on Earth. Since the small primordial polyatomic units are scattered throughout the Universe, it is indeed difficult to imagine in all this immensity of space that the Earth is the only planet where environmental conditions were sufficient to trigger the process of life.

No exceptional factor dictated improvements in the organization of matter. Usually, the reactions of the first small molecules were in the direction of synthesizing prebiotic compounds that were more complex and stable; never in the reverse direction, apparently! For example, we know that hydrocyanic acid can, with the participation of ammonia, generate adenine but it is difficult to imagine this happening the other way around! Generally, chemical evolution is a one-way street; it cannot go backwards any more than can a ratchet mechanism. The newly created and more complex structures, if they are stable enough, are established and persist; when they do not, they simply vanish. As far as the later phases of the development of matter are concerned, thermodynamics provides a good explanation of how systems that are not in a state of equilibrium can reach higher degrees of organization. We have also seen that a process as essential as self-replication is found in certain "inanimate" structures such as clays or polynucleotides. In the latter case, an explanation has now been worked out as to how a process with a primitive origin, low efficiency, and poor reliability can rise to the level of a self-reproducing program-based system and thus perpetuate a lineage of structures copied from generation to generation with good fidelity. The infrequent program changes offer the possibility of advan tageous diversification: new organisms undergo selection based on their ability to reproduce; if they are better adapted to the environment, they become more prolific.

So we understand the reasons for the ongoing evolution of matter toward more complex and more highly perfected forms. Progress is not accomplished "actively," guided by some predetermined "plan" or "goal," but by the elimination of the least well-adapted structures by what could be called an upward leveling. Natural selection applies not only to living organisms but also to molecules, even small ones: any chemical entity exists only if the conditions in its environment allow it.

The chemical development of matter was easier in that Nature did not have to show a great deal of creativity in this area. At the different stages of molecular organization, the first structures to appear were relatively simple and few in number. But it is the highly varied combination of these basic elements that led to the development of new structures and new functions of striking diversity. The advent of life, this remarkable event which occurred in the Universe, thus cannot even be considered a miracle of chemistry. In fact, chemistry did no more than to produce molecules whose appearance at a given point in time and under specific conditions obeyed simple laws. The subsequent emergence of "living" structures of infinite complexity appeared to be more like a recombination phenomenon within the framework of a progressive self-organization of matter tending to higher degrees of order. We find the same process at work at the different stages of organization: the combination of a restricted number of elements (amino acids, nucleotides, exons, etc.) led to a number, often a huge number, of more complex structures.

In principle, the possibilities of this combinatorial strategy are limitless: we have cited above several examples attesting to the hyperastronomical number of structures it is capable of producing. This powerful strategy created not only the biosphere but, by combining physical and chemical interactions of an enormous number of neurons together, was even capable of establishing a supreme degree of organization of matter, even to the point of creating thought and awareness.

There is no need to imagine that the creation of the world and life was the work of God; on the contrary, the notion of God emerged only with the advent of thought—unless we wish to ascribe divine power to the set of physical and chemical laws that governed the organization of matter from the origin of the Universe to the appearance on the scene of humankind!

These ideas may seem too reductionist and jolt those who believe that life cannot be boiled down to a purely physical and chemical scheme. In their eyes, additional explanations are necessary, particularly as concerns the functioning of the higher levels of organization of life. "Biology can neither be reduced to physics nor do without it," said François Jacob. It should be pointed out, however, that despite the considerable advances in the study of biological systems to date, it was not necessary either to invent or to implement any physical or chemical law of a new kind to account for life.

Thus far, it has been possible to account for all observations by known laws—the same laws that govern simple chemical phenomena. Let us not, in our own sphere, repeat the errors of the old "vitalism," and let us be prepared to consider the living organism as a chemical machine that has attained an extraordinary degree of complexity. At most, chemistry carried to such a high degree of sophistication could merit being called "living chemistry."

Such a materialistic affirmation may seem highly venturesome at the present stage of our knowledge. But it is becoming less and less so with time as science progresses. Phenomena that formerly seemed beyond understanding by reason of their complexity are now amenable to satisfactory explanations. Think of the formidable challenge that the mechanism of heredity represented for biology for such a long time! Today we know the main molecular foundations of heredity. Likewise, we are now beginning to lift the veil that until quite recently concealed so extraordinarily complicated and apparently enigmatic a process as cell differentiation, leading to the formation of a complex organism from a single cell, the fertilized egg.

Today we are witnessing a similar phenomenon: dazzled and dizzied by the intricacies of the functioning of the brain, certain biologists have invoked Gödel's principle, according to which it is not possible for a logical system to describe itself, and have stated accordingly that it will never be possible to penetrate the mysteries of the molecular biology of the brain's higher activities (memory, thought, feelings, etc.). This may be true. But our entire history is replete with examples of mental barriers that were erected successively at various times between what was accessible to our understanding and what at that point was beyond us. Metaphysical philosophies, mystical dogmas, and religions sprang from the existence of these barriers, on one side of them the logical and explainable phenomena of material essence and on the other side, those of divine essence. Yet, progress in science is shifting these frontiers constantly.

In any event, the interpretation of biological phenomena on the basis of chemistry will certainly increase in scope and in the future become a new scientific discipline. This deeper molecular biology, going beyond the molecular biology of today, offers a fascinating field to researchers. It goes without saying that the very legitimate intellectual satisfactions that can be derived from future discoveries in this field will not be the only benefit of this basic research. We may predict that better knowledge at the microscopic level of how organisms function will have beneficial repercussions on the development of medicine, agriculture, ecology, and, in general, all the numerous and various fields in which biology influences the life of society.

BIBLIOGRAPHY

BRESLOW, R., *Mécanismes des réactions organiques* [Mechanisms of organic reactions],* Édiscience, Paris, 1970.

CAIRNS-SMITH, A. G., *Genetic Takeover and the Mineral Origins of Life*, Cambridge University Press, Cambridge, 1982.

DEBRU, C., *L'Esprit des protéines* [The spirit of proteins],* Hermann, 1983.

GROS, F., *Les Secrets du gène* [The secrets of the gene],* Odile Jacob, 1986.

JACOB, F., *La Logique du vivant* [The logic of living things],* Gallimard, 1970.

MONOD, J., *Le Hasard et la Nécessité* [Chance and necessity],* Seuil, 1970.

OPARIN, A. I., *L'Origine de la vie sur la Terre* [The origin of life on Earth],* Masson et Cie, Paris, 1965.

PRIGOGINE, I., and STENGERS, I., *La Nouvelle Alliance* [The new alliance],* Gallimard, 1979.

* These references have not been published in English.